United States Department of Agriculture

Economic
Research
Service

Economic
Research
Report
Number 217

October 2016

Farm Profits and Adoption of Precision Agriculture

David Schimmelpfennig

United States Department of Agriculture

Economic Research Service
www.ers.usda.gov

Access this report online:

www.ers.usda.gov/publications/err-economic-research-report/err217

Download the charts contained in this report:

- Go to the report's index page www.ers.usda.gov/publications/ err-economic-research-report/err217
- Click on the bulleted item "Download err217.zip"
- Open the chart you want, then save it to your computer

Recommended citation format for this publication:

Schimmelpfennig, David. *Farm Profits and Adoption of Precision Agriculture,* ERR-217, U.S. Department of Agriculture, Economic Research Service, October 2016.

To ensure the quality of its research reports and satisfy government-wide standards, ERS requires that all re-search reports with substantively new material be reviewed by qualified technical research peers. This technical peer review process, coordinated by ERS' Peer Review Coordinating Council, allows experts who possess the technical background, perspective, and expertise to provide an objective and meaningful assessment of the output's substantive content and clarity of communication during the publication's review. For more information on the Agency's peer review process, go to: http://www.ers.usda.gov/about-ers/peer-reviews.aspx

United States Department of Agriculture

Economic Research Service

Economic Research Report Number 217

October 2016

Farm Profits and Adoption of Precision Agriculture

David Schimmelpfennig

Abstract

Precision agriculture (PA) and its suite of information technologies—such as soil and yield mapping using a global positioning system (GPS), GPS tractor guidance systems, and variable-rate input application—allow farm operators to fine-tune their production practices. Access to detailed, within-field information can decrease input costs and increase yields. USDA's Agricultural Resource Management Survey shows that these PA technologies were used on roughly 30 to 50 percent of U.S. corn and soybean acres in 2010-12. Previous studies suggest that use of PA is associated with higher profits under certain conditions, but aggregate estimates of these gains have not been available. In this report, a treatment-effects model is developed to estimate factors associated with PA technology adoption rates and the impacts of adoption on profits. Labor and machinery used in production and certain farm characteristics, like farm size, are associated with adoption as well as with two profit measures, net returns and operating profits. The impact of these PA technologies on profits for U.S. corn producers is positive, but small.

Keywords: Crop production information technologies, precision agriculture, variable-rate technology, soil tests, global positioning system maps, guidance systems.

Acknowledgments

The author would like to thank branch chief, Jim MacDonald, and the rest of the management team, including Pat Sullivan, Mark Jekanowski, Robert Gibbs, Cindy Nickerson, and Marca Weinberg, U.S. Department of Agriculture (USDA), Economic Research Service; as well as Rachael Brown (formerly with USDA/ERS) and Robert Ebel (now with USDA's Risk Management Agency). He would also like to thank the following for technical peer reviews: David Bullock (University of Illinois at Urbana-Champaign), Jeremy Foltz (University of Wisconsin-Madison), and three reviewers who requested anonymity, two from USDA's Natural Resources Conservation Service. Thanks also to Dale Simms and Curtia Taylor, USDA/ERS, for editorial and design services.

Contents

Find the full report at *www.ers.usda.gov/publications/err-economic-research-report/err217*

Farm Profits and Adoption of Precision Agriculture

David Schimmelpfennig

What Is the Issue?

How and whether farm managers decide to adopt new technologies is complex, but most account for the full costs and benefits of the proposed investment. Precision agriculture (PA) technologies require a significant investment of capital and time, but may offer cost savings and higher yields through more precise management of inputs. Until the early 2000s, the adoption rate of different PA technologies varied up to 22 percent across major U.S. field crops. After that time, adoption of some technologies began to outpace others. Yield mapping via Global Positioning System (GPS) grew faster for corn and soybeans than for other crops, while adoption of soil mapping varied substantially across crops. Tractor guidance systems have grown faster than variable-rate input application for all major field crops over the last 10 years.

This study investigates recent trends in PA adoption, the production practices and farm characteristics associated with adoption, and whether adoption is associated with greater profitability.

What Did the Study Find?

This report examines adoption rates for three types of PA technologies: (1) GPS-based mapping systems (including yield monitors and soil/yield mapping); (2) guidance or auto-steer systems; and (3) variable-rate technology (VRT) for applying inputs.

- **Adoption rates vary significantly across PA technologies:** Yield monitors that produce the data for GPS-based mapping are the most widely adopted, used on about half of all corn (2010) and soybean (2012) farms, while guidance or auto-steer systems are used on about a third of those farms and GPS-based yield mapping on a quarter. Soil mapping using GPS coordinates and VRT are used on 16 to 26 percent of these farms.

- **The largest corn farms, over 2,900 acres, have double the PA adoption rates of all farms:** 70-80 percent of large farms use mapping, about 80 percent use guidance systems, and 30-40 percent use VRT.

ERS is a primary source of economic research and analysis from the U.S. Department of Agriculture, providing timely information on economic and policy issues related to agriculture, food, the environment,and rural America.

www.ers.usda.gov

— The share of all corn and soybean *acres* on which PA technologies are used tends to be higher than the share of *farms*, implying that larger farms are more likely to adopt these technologies. Yield mapping is used on about 40 percent of U.S. corn and soybean acres, GPS soil maps on about 30 percent, guidance on over 50 percent, and VRT on 28-34 percent of acres.

- **PA technology adoption and farm size both influence production costs on corn farms:**

  — Hired labor costs are 60 to 70 percent lower with any of the three PA technologies on small corn farms (140-400 cropland acres), while hired labor costs are higher on large farms that have adopted precision mapping and guidance. The additional use of hired labor on larger farms may be for information management and field operation specialists that can help implement PA technologies. Larger farms have higher expenses for other inputs that these specialists can help control using PA. Custom service expenses are higher with mapping and guidance on both large and small corn farms under all three PA technologies. However, custom operation costs are five times larger, in percentage terms, on small farms than on large farms.

- **Statistical analysis finds that several production inputs and practices are associated, both positively and negatively, with adoption of PA technologies on corn farms:**

  — Non-GPS-based soil testing increases the adoption of all three PA technologies.

  — Higher levels of unpaid labor and higher yield goals, representing the farmer's self-reported yield potential, have a negative effect on PA adoption. Unpaid labor is a large, fixed overhead expense that may reduce the flexibility to adopt PA technologies. When yield goals are higher, farmers may already be close to the production potential for their land, whereas farmers with lower yield goals may be using the technologies to try to raise yields on land known to be less productive.

  — A bigger stock of machinery on corn farms has a negative effect on VRT adoption, possibly because of higher overhead costs, and less flexibility in taking on new capital outlays.

- **All three technologies have small positive impacts on both net returns (including overhead expenses) and operating profits for a U.S. corn farm of average size:**

  — GPS mapping shows the largest estimated impact among PA technologies, with an increase in operating profit of almost 3 percent on corn farms. The impact of mapping on net returns is almost 2 percent.

  — Guidance systems raise operating profit on corn farms by an estimated 2.5 percent and net returns by 1.5 percent.

  — Variable-rate technology (VRT) raises both operating profit and net returns on corn farms by an estimated 1.1 percent.

- **Corn and soybeans have had higher shares of acreage using yield mapping than other crops, but use of yield maps has increased for peanuts, rice, and spring wheat as well.**

Adoption of yield mapping (by crop)

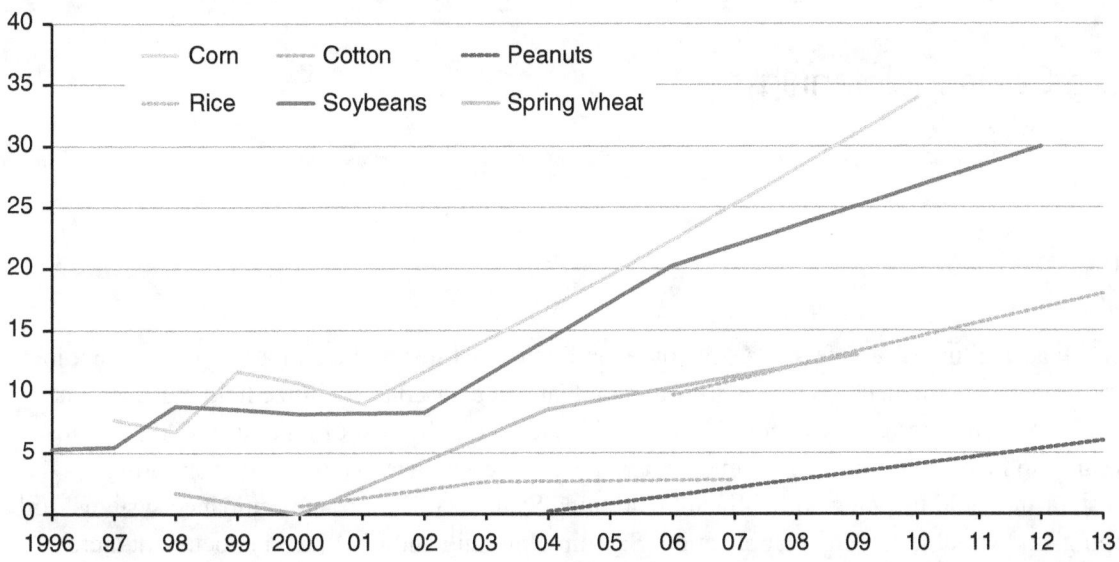

Percent of crop planted acres

Legend: Corn · Cotton · Peanuts · Rice · Soybeans · Spring wheat

Source: USDA, Economic Research Service estimates using data from the Agricultural Resource Management Survey (ARMS) Phase II.

How Was the Study Conducted?

This study uses national data on U.S. field crop production between 1996 and 2013 (the latest available) from the Agricultural Resource Management Survey (ARMS) of field crop producers, jointly administered by USDA's National Agricultural Statistics Service (NASS) and ERS. The survey data allow examination of detailed field-level production and financial information for a large sample of farms. Use of three PA technologies—information mapping, guidance systems, and VRT—is examined for each farm. An econometric model is estimated that controls for selection bias since large corn farms tend to be both more profitable and early technology adopters.

Farm Profits and Adoption of Precision Agriculture

David Schimmelpfennig

Introduction

Precision agriculture (PA) refers to a suite of technologies that may reduce input costs by providing the farm operator with detailed spatial information that can be used to optimize field management practices (National Research Council, 1997). Besides reducing the cost of input use, PA technologies can also increase yields. These benefits derive from the efficient use of yield-monitoring harvesters and yield mapping with Global Positioning Systems (GPS), tractor guidance systems, soil mapping, and variable-rate input application. Schimmelpfennig and Ebel (2011) document increasing adoption of these technologies from 1997 to 2005. While these are the most popular PA technologies and could be leading to higher profits for field crop producers, adoption rates are generally less than 50 percent, with variable-rate technology (VRT) lagging the others.

This report uses data from the Agricultural Resource Management Survey (ARMS) to identify factors associated with adopting PA technologies on field crop farms and to examine whether farmers who use these technologies have higher operating profits and net returns (which includes overhead expenses) than similar farmers who do not.

Even with a technology that increases farm profits, adoption is often slow at first and faster later, owing to farm characteristics and the learning required to integrate new technology with existing practices. A 2012 review of large-scale agriculture in countries with long-term PA experience found that many farm characteristics, beyond input use and output level, affected adoption of PA technologies and that the impacts of PA adoption on profits were mixed (Tey and Brindal, 2012). In particular, the impact of PA technologies on profits for U.S. corn producers was found to be small but positive, which may explain the slow but steady growth in adoption.

However, simple comparisons of farm profit between adopters and non-adopters of PA technologies can be misleading because some of the characteristics impacting profit—such as farm size, since larger farms tend to be more profitable—may also influence whether a farmer adopts the technology. For this reason, more rigorous statistical analysis of technology adoption is used to examine the relationship between adoption and profit.

Measuring Farm Profit

This report uses two key measures of farm profit: ***operating profits***, or revenue less production costs, and ***net returns*** that includes overhead in costs. The data needed to calculate these measures come from the Agricultural Resource Management Survey (ARMS), which provides detailed field-level production information for a nationally representative sample of farms as well as financial information for each farm.

PA technologies can reduce operating costs by preventing farmers from overapplying inputs. Even if input use and operating costs increase under PA, yields can grow enough to increase operating profits. Both effects would be reflected in operating costs and profits. The capital expenditures needed to implement PA technologies could raise overhead costs but also enable operators to substitute capital and labor for operating inputs. In this case, operating costs would fall and operating profits would rise, but the effect of PA on capital and labor use must be captured in order to evaluate the impact on net returns. Because PA technologies can affect farm economics through multiple channels, we need to evaluate both net returns and operating profits.

Descriptions of Precision Agricultural Technologies

Yield monitors have not changed much since they were first used on combines in the 1990s, but the ease and functionality of yield/soil data maps, guidance systems, and variable-rate technology have slowly evolved.

Computer Mapping

Farmers use GPS-based computer mapping of yield and soil data to help customize crop management across and within fields. When areas in a farmer's fields are lower yielding or have less productive soil, the farmer either adds inputs to raise yields (and costs) or reduces inputs on areas that are lower yielding and less likely to be profitable. The specific technologies needed to gather yield and soil information are both GPS-based, but differ slightly.

For farm *yield* data, harvester-mounted yield monitors gather data in the grain elevator of the combine. As paddles in the elevator rotate and eject the harvested corn to a waiting truck, the grain bounces off a load cell that measures the corn mass-flow that is converted to an electrical signal captured on a flash drive in the chute. The device collects simultaneous GPS coordinates and the data can be mapped to show within-field yield variability by using color-coded data points within the geographic boundaries of a farmer's field. With ever-expanding storage space on flash drives, combine manufacturers have programmed the drives to retain many years of yield data, allowing multi-year yield comparisons on maps and requiring just a computer and expertise in interpretation, making this technology less costly than guidance systems and variable-rate technology (VRT).

Soil maps are often created for soil types, soil nitrate levels, and pH level. These maps are generally created by plotting soil characteristics data on personal computers with larger screens and sufficient resolution to visualize field details.[1] These maps may be used to determine the need for VRT, to determine optimal seeding rates, or to justify alternative land uses like soil conservation, hay, or pasture.

The data to create a soils map may come from public sources, which can be augmented with data from lab results of core soil samples or from onsite electrical conductivity tests. USDA's National Agriculture Imagery Program (NAIP), for example, acquires aerial imagery during the agricultural growing season and makes these data available to the public. USDA's National Cooperative Soil Survey is another public data source. Promising technologies to remotely develop and map soils data from unmanned aerial vehicles (drones) and aircraft are in the proof-of-concept stage and product development.

Compared to yield monitor data that are produced almost continuously from combines, data from soil testing technologies are sparse and intermittent. Producers often consult input providers and co-ops for analysis and mapping of yield and soils data, but several mobile phone application developers offer mapping/soil testing services as well.[2]

---

[1] Yield and soil data are often combined as overlays on the same map.

[2] Rudimentary soil maps are often developed at the 5-acre level and paired with topographical information, along with yield information, to help guide VRT application equipment (PrecisionAg Buyer's Guide, 2013).

Guidance Systems

Adopting GPS-guided or auto-steered combines and tractors can reduce operator errors by determining precise field locations (that are often difficult to determine accurately by sight) and compensating for operator fatigue. Field operators using guidance systems have timely, accurate coordinates accessible from a screen in the cab. Guidance systems save money by reducing over- and under-application of sprays and better aligning the seeding of field crop rows. This enables harvesting with fewer problems from densely spaced corn stalks that can cause combines to malfunction. Guidance systems also free operators from steering, allowing them to potentially monitor several PA systems at once.

Guidance systems are most often used on tractors. Combine harvesters are also being fitted with guidance systems to help keep equipment precisely on corn and soybean rows.[3] While these systems are not standard equipment on new tractors, most are guidance systems-ready, requiring additional investment in a GPS receiver with the level of spatial resolution desired.

Variable-Rate Technology

Customized seeding and application of fertilizer, chemicals, and pesticides is accomplished with machinery attachments that can vary the rate of application from GPS controls in the cabs of tractors. Geolocated data from yield and soil maps or from guidance systems can be used to pre-program application equipment to apply desired levels of inputs or to seed at pre-determined rates at different locations in a farmer's field. Controllers adjust the levels of inputs coming from each nozzle or feeder on command from a computer program that uses the geo-referenced data points. This system has recently been expanded to allow different types of hybrid corn seeds to be planted at different locations in a farmer's field with a single pass of the tractor, a technology known as prescription planting.

The capital cost of farm implements equipped with VRT capabilities is fairly high, especially when specialized machinery with integrated sprayer or seeding equipment must be scrapped. For this reason, many producers—particularly on smaller operations—have opted to hire service providers when choosing VRT. Only 21 percent of the PA studies reviewed by Griffin et al. (2004) included human capital costs, but operator time and effort was found to be a substantial cost for VRT and a likely reason for outsourcing the service.

---

[3] Erickson et al. (2013) used the Purdue University survey of full-service input suppliers to track the popularity of guidance systems among service providers and found that 82 percent were using some form of guidance systems with their services.

Adoption of Precision Agricultural Technologies

The 2010 ARMS provides data to measure recent adoption rates of the main PA technologies on corn farms (figure 1). Nearly half of corn farms used yield monitors. Even though yield mapping uses data from yield monitors, mapping was less commonly used (25 percent of corn farms). Guidance systems are the second-most frequently adopted PA technology, at 29 percent of corn farms. GPS soil mapping had been adopted by 19 percent of corn farms by 2010, as had VRT use. Yield mapping may also be used for VRT programming.

PA technologies have been applied to higher percentages of corn acres than corn farms, implying that larger farms are more likely to adopt these technologies (figure 1). The difference in share of farms and acres is particularly large for guidance systems, 29 percent versus 54 percent. VRT adoption appears to depend less than the other technologies on farm size. VRT adoption may also depend on whether a farm has significant soil and yield variability, a hypothesis for future research.

PA adoption rates are similar on U.S. soybean farms (figure 2, based on 2012 ARMS), which is plausible since about half of all corn farms annually rotate corn with soybeans. The largest difference in adoption rates between corn in 2010 and soybeans in 2012 is for VRT, which is 7 percent higher with soybeans. This difference could be due to characteristics of soybean growing areas where corn is not grown, or production practices used on soybeans but not corn. The difference is also consistent with increasing adoption of PA technologies over time, as soybean farmers were surveyed 2 years after corn farmers were. As for corn, larger percentages of cropland acres (farm size) adopt all of the technologies, and VRT farms and cropland have the smallest difference in adoption percentage.

ARMS surveys track the adoption of PA technologies across six different crops surveyed in selected years: corn (1996-2000, 2005, 2010); cotton (1996-2000, 2003, 2007); peanuts (1999, 2004, 2013);

Figure 1
Adoption of PA technologies on corn farms, 2010

Percent of corn farms/cropland

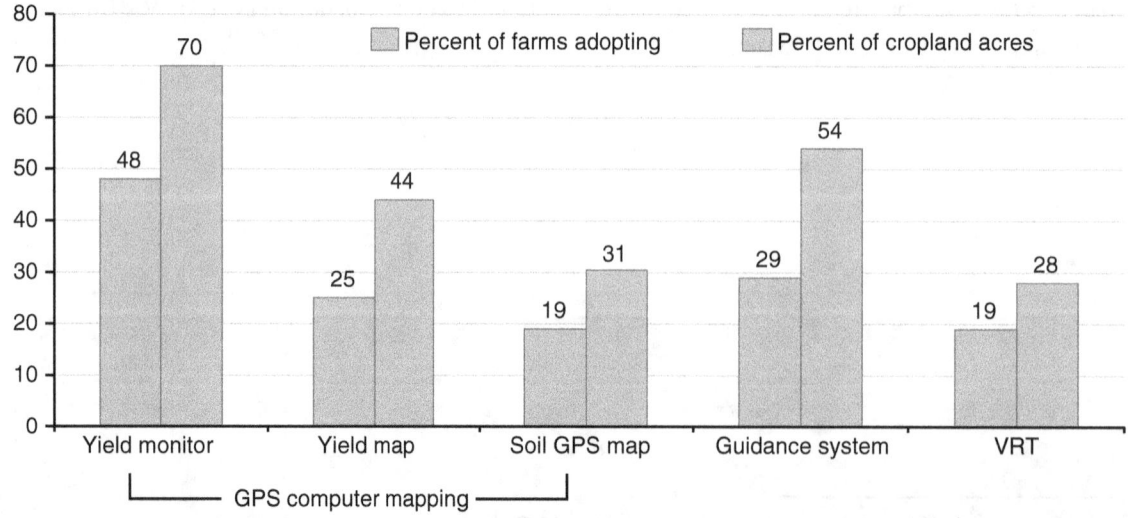

GPS = Global Positioning System. PA = precision agriculture.
Source: USDA Economic Research Service estimates using data from the Agricultural Resource Management Survey (ARMS) Phase II.

Figure 2

Adoption of PA technologies on soybean farms, 2012

Percent of soybean farms/cropland

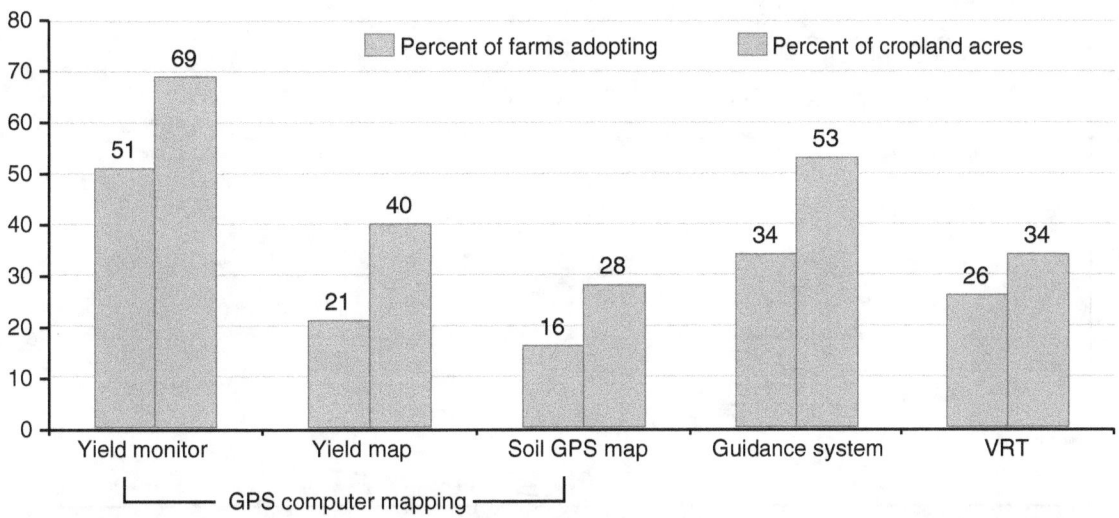

GPS = Global Positioning System. PA = precision agriculture.
Source: USDA, Economic Research Service estimates using data from the Agricultural Resource Management Survey (ARMS) Phase II.

rice (2006, 2013); soybeans (1996-2000, 2002, 2006, 2012); and spring wheat (1996-1998, 2000, 2004, 2009). Adoption has generally risen, but at different rates.

Corn and soybeans have had higher shares of acreage using yield mapping than the other four crops, but use of yield maps has increased for peanuts, rice, and spring wheat as well. Yield mapping on corn and soybean crops grew from just under 10 percent in 2001-02 to 30 percent or higher in 2010-12 (figure 3). The data for figures 3-6 are shown in the box, "ARMS Data."

Adoption of soil mapping (figure 4) shows a different pattern, with all crops except peanuts and rice approaching current levels (10-20 percent) by 2000. Peanut adoption of soil mapping was low until 2004, but then eclipsed the other crops in 2013 at almost 25 percent.

Guidance system adoption (figure 5) shows a striking similarity in rate of increase across all surveyed crops. By 2013, guidance was used on 45-50 percent of all crops except cotton. (Cotton was surveyed again in 2015 and may show a similar jump in adoption.)

VRT (figure 6) has not shown the consistent increases in adoption that guidance has, but by 2012, corn, soybeans, and rice were all above 20 percent.

Precision Agriculture Technologies Often Used in Tandem

Precision agriculture technologies are often adopted in different combinations. Adoption percentages of each technology alone and in combination with others show that GPS mapping is more often adopted alone (17.2 percent of corn farms) than in combination with other PA technologies; it is adopted with guidance on 5.7 percent of farms, with VRT on 4.3 percent of farms, and as a combination of all three on 3.8 percent of farms. (These four percentages equal the total adoption share for GPS mapping of 31 percent in 2010; see figure 1.)

Figure 3
Adoption of yield mapping (by crop)

Percent of crop planted acres

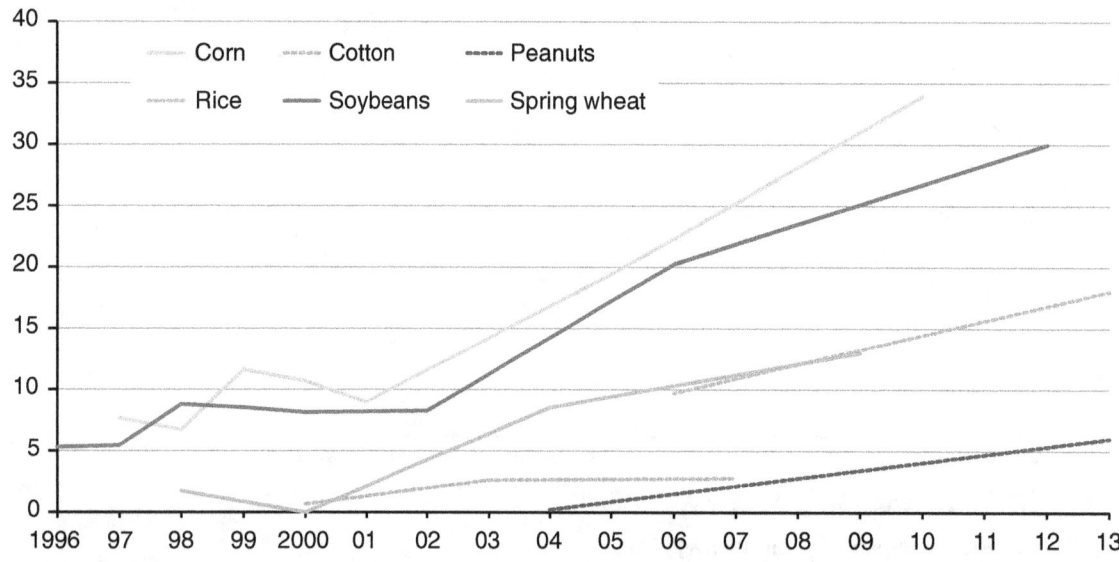

Source: USDA, Economic Research Service estimates using data from the Agricultural Resource Management Survey (ARMS) Phase II.

Figure 4
Adoption of GPS soil mapping (by crop)

Percent of crop planted acres

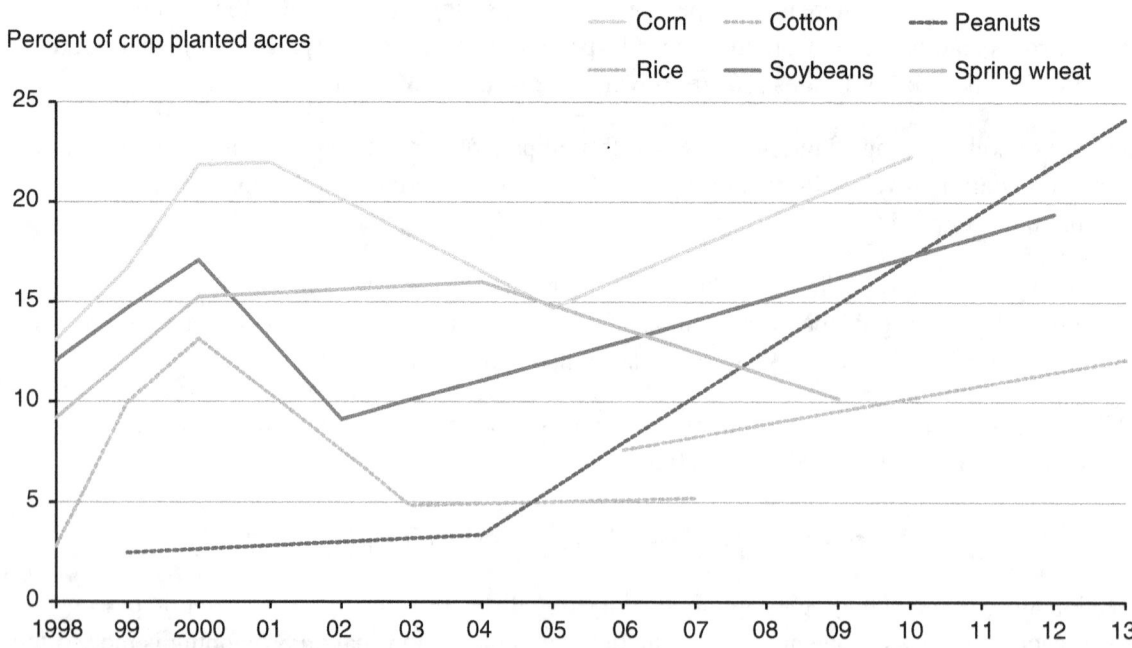

Note: Global Positioning System (GPS).
Source: USDA, Economic Research Service estimates using data from the Agricultural Resource Management Survey (ARMS) Phase II.

Figure 5
Adoption of guidance systems (by crop)

Percent of crop planted acres

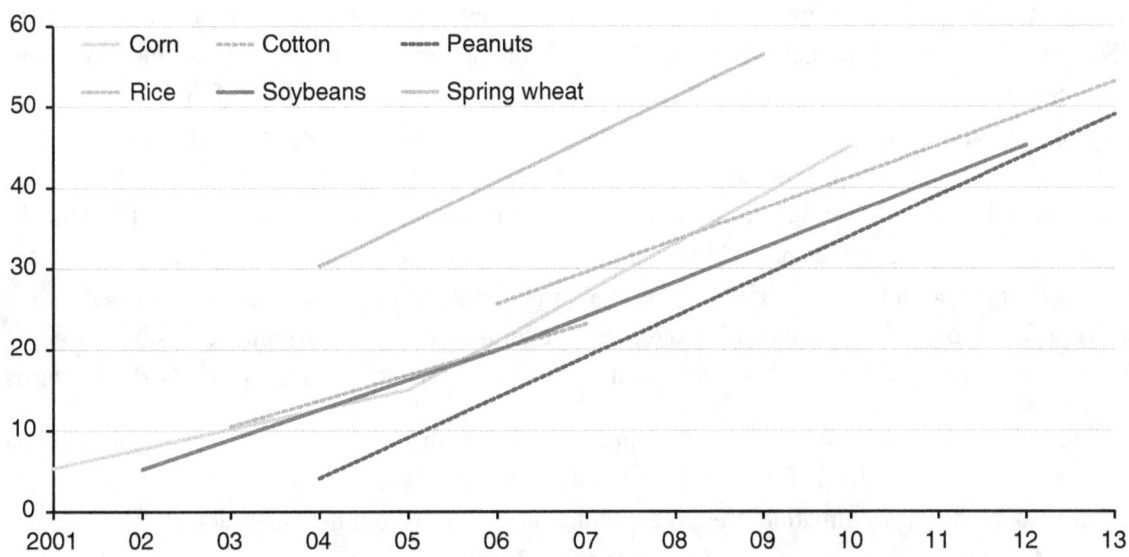

Source: USDA, Economic Research Service estimates using data from the Agricultural Resource Management Survey (ARMS) Phase II.

Figure 6
Adoption of variable-rate application technology (VRT) by crop

Percent of crop planted acres

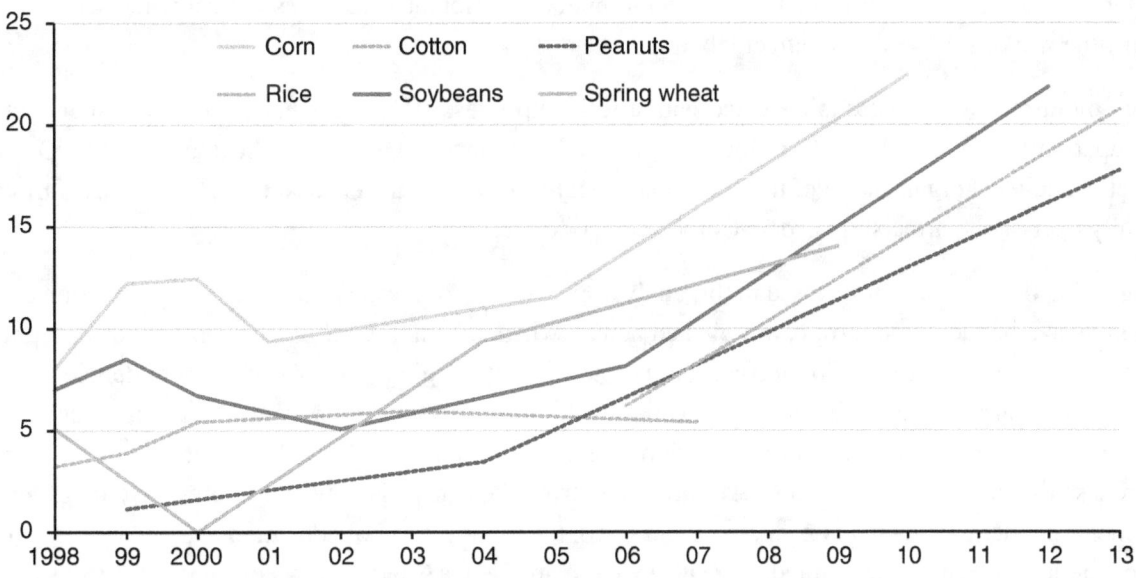

Source: USDA, Economic Research Service estimates using data from the Agricultural Resource Management Survey (ARMS) Phase II.

Box: ARMS Data

The data used in this report come from the Agricultural Resource Management Survey (ARMS) of field crop producers. ARMS is a large annual farm survey managed jointly by USDA's Economic Research Service and National Agricultural Statistics Service. Phase III of the survey collects data on farm production, financial outcomes, and farm household attributes for a sample representing all farms in the contiguous 48 States. Separately, the survey collects field-level information on production practices for one to two selected crops in Phase II; producers of those crops are also contacted for the farm and farm household phase of the survey. Detailed data provide crucial information on inputs used for agricultural production like machinery and labor, as well as the use of precision technologies including GPS mapping, guidance systems, and variable-rate application (VRT). Technology adoption is measured as percent of farms answering yes/no to individual technology-use questions weighted by a deflation factor for overrepresentation of large farms in the sample. Phase II of ARMS also collects data on yields obtained, operating profit, and net returns.

Operating profit is the gross value of production, including corn as grain and as silage, minus variable production costs. Crop production in bushels and costs in dollars are often converted to a per-acre basis for comparison and empirical estimation. Variable production costs are the sum of chemical, custom operation, electricity, fertilizer, fuel, interest (on operating capital), irrigation water, pesticide, repair, and seed costs. Custom service costs are collected for specific tasks such as custom seeding or harvesting that are paid by task rather than by hour. Net profits or total net returns includes all allocated overhead items in costs and is smaller (or more negative) than operating profit (table 5). Allocated overhead includes hired labor, the opportunity cost of unpaid labor, machinery and equipment capital service flows, the opportunity cost of land measured as its rental rate, taxes and insurance, and other general farm overhead. Hourly labor is collected in the ARMS for time spent operating machinery; scouting for weeds, insects, and diseases; and other manual work. These labor hours are multiplied by different wage rates for part-time/seasonal and full-time workers, as well as contract laborers.

Data on farm-level finances is collected with a followup Phase III questionnaire on corn farmer primary occupation (Occupation), off-farm income, years of experience, age, farm debt-to-asset ratio (Debt-to-assets), and legal organization of the operation (Legal-org). Farmers were asked to classify their farm as a family operation, partnership, C or S-Corporation, or estate/trust.

The Phase II 2010 corn survey used in this analysis provided 1,360 observations on corn production details like the size of each corn farm (Corn planted acres), whether a (non-GPS based) soil test for phosphorus or nitrogen was performed (Soil testing, but not necessarily mapping), or if leaf nutrient deficiency tests were done. Farmers were also asked if genetically modified seed varieties were planted on a majority of the acreage (GMO seeds), if no-till operations were performed (No-till used), and if a tractor had been purchased since 2005 (New tractor). Also available from the dataset is how many acres were irrigated, the farmer's per-acre yield goal at planting (Yield goal), the highest level of education attained by the farmer, and whether a consultant was hired to create a yield map. Tables 6 and 7 show descriptive statistics (mean, standard deviation, min and max) for the data used in the estimations.

The adoption percentages (of crop planted acres) used on figures 3-6 are available at http://www.ers.usda.gov/data-products/arms-farm-financial-and-crop-production-practices/tailored-reports-farm-structure-and-finance.aspx.

Guidance systems are also more often adopted alone (16.1 percent of corn farms); they were used in combination with GPS mapping (5.7 percent) and VRT (3.4 percent), with all three together on 3.8 percent of farms. (These four percentages equal the total adoption share for guidance systems of 29 percent in 2010.) See box, "Key Precision Ag Technologies and Information Flows," for more discussion of information flows from data collection to mapping to data use.

Variable-rate technology, on the other hand, is more often adopted in combination with other PA technologies than alone: VRT alone is adopted on 7.5 percent of corn farms and in combination with GPS mapping (4.3 percent) and guidance (3.4 percent), with all three used together on 3.8 percent of farms. (These four percentages equal the total adoption share for VRT of 19 percent in 2010).

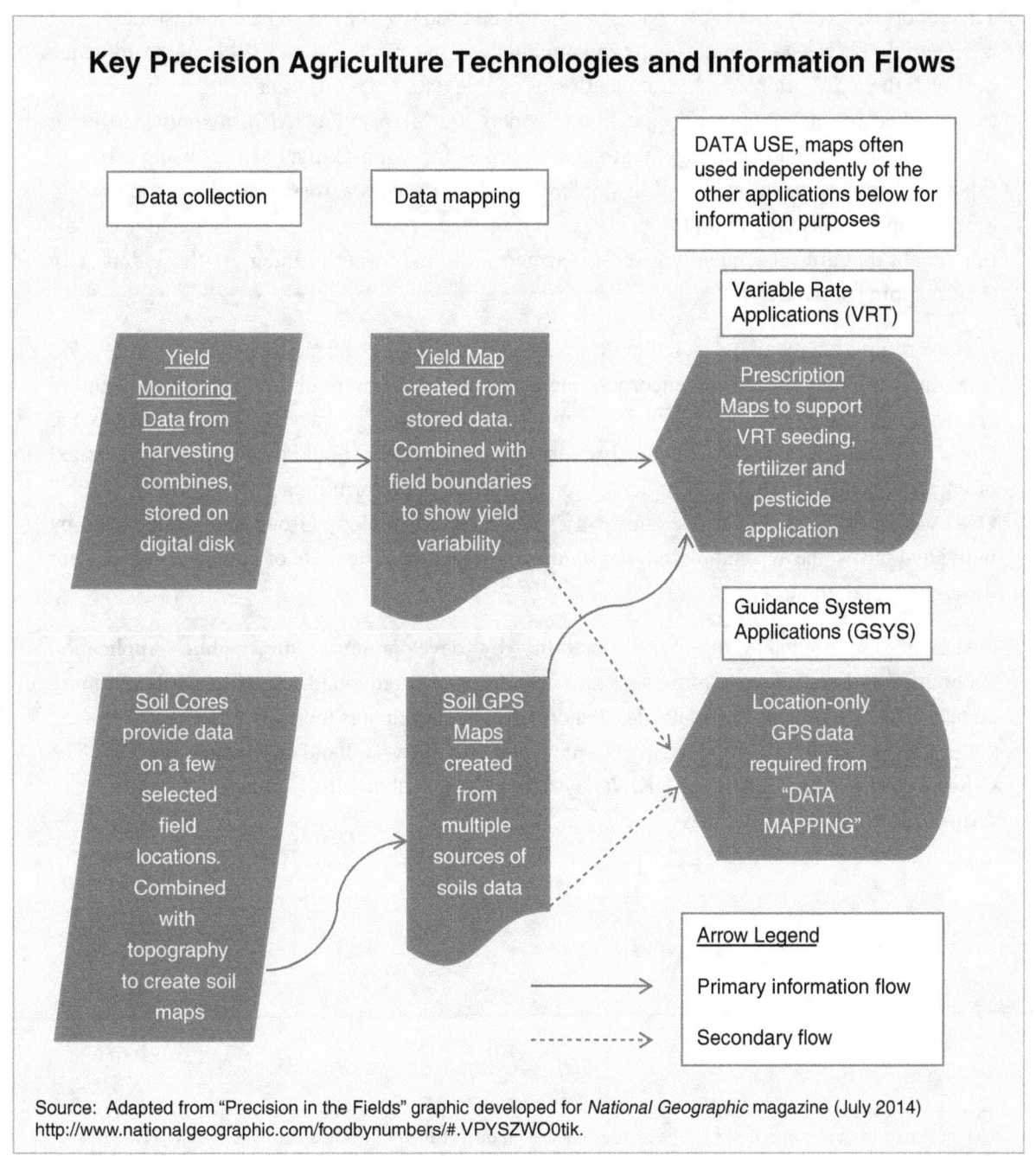

Key Precision Agriculture Technologies and Information Flows

Source: Adapted from "Precision in the Fields" graphic developed for *National Geographic* magazine (July 2014) http://www.nationalgeographic.com/foodbynumbers/#.VPYSZWO0tik.

Farm Profits and Adoption of Precision Agriculture, ERR-217
Economic Research Service/USDA

Data privacy and security issues may influence adoption of different combinations of PA tools. Several agricultural input companies, including Monsanto and John Deere, are connecting data from different technologies in one farmer's fields by offering platforms that can store data from various sources.[4] Farmers may be wary of these products because, even though their privacy is protected and their data anonymous, the data are still linked to their GPS coordinates and subsequent use of data is beyond the farmer's control (see box, "Data Platforms," for more discussion of data privacy and security issues).

Data Platforms

This report describes how PA technologies might be used together in some circumstances. The connectivity between computer maps and applications like guidance and VRT might influence how well the applications work. Data platforms enable data from different sources on one farm to be linked in one computer application. Farmers are often interested in mapping different data, like soil and yield data, together to help improve decisionmaking. Farmers using different types of data have been interested in standardizing data access across desktop computers and mobile applications taken into the field. Custom service providers often produce and use additional data that are subsequently given to farmers, who may want to integrate those data with their mapping software.

A consortium of companies, including Monsanto and Deere, has formed AgGateway to develop best practices and standards for incorporating data privacy into agricultural operations. Another organization concerned with the interoperability, security, and privacy of farmers' data is the Open Ag Data Alliance (OADA), which asserts that each farmer should own the "data generated or entered by the farmer, their employees, or by machines performing activities on their farm." OADA is an extension of Purdue University's Open Ag Technology Group and is supported by individual farms, the agriculture industry, and USDA's National Institute of Food and Agriculture through several grants.

OADA seeks to enable data interoperability by developing secure, public Application Programming Interfaces (APIs) so that farmers can use trusted cloud providers while retaining control over who can use their data. The list of industry participants includes AgReliant Genetics, CNH (the parent for Case International Harvester and New Holland machinery brands), The Climate Corporation, GROWMARK, Valley® Irrigation, Wilbur-Ellis Company (AgVerdict platform), and WinField (CROPLAN seeds).

---

[4] Monsanto offers a platform for its seed and weed control products called Integrated Farming Systems (IFS) that will include weather data from The Climate Corporation in precision planting and other agronomic recommendations. DuPont Pioneer and John Deere are collaborating to link precision agronomy software with John Deere wireless data transfer from machinery called JDLink™.

Adoption of Precision Agriculture Technology and Farm Size

Larger farms are most likely to adopt PA technologies (table 1; Fernandez-Cornejo et al., 2002). Here, farm size is defined by acres of cropland operated, whether planted to corn or another crop.[5] The highest adoption rates for all three technologies in 2010 were on farms over 3,800 acres. Computer mapping (both GPS soil and yield mapping) and guidance adoption were adopted on over 40 percent of all farms above 1,300 acres. VRT adoption (the costliest of the three) is lower than for the other technologies on all farm sizes. VRT adoption is more prevalent on farms over 1,700 acres than on farms under 1,700 acres.

Table 1

Precision agriculture adoption is higher on larger corn farms (2010)

Cropland acres*	GPS soil/yield mapping	Guidance system	VRT
	Percent of farms adopting each technology		
Less than 600 acres	12%	12%	12%
600 – 1,000 acres	34%	24%	20%
1,000 – 1,300 acres	39%	33%	18%
1,300 – 1,700 acres	50%	40%	23%
1,700 – 2,200 acres	54%	60%	32%
2,200 – 2,900 acres	49%	60%	32%
2,900 – 3,800 acres	67%	78%	29%
Over 3,800 acres	80%	84%	40%

*Note: Cropland acres are all farm acres planted to any crop whether owned or rented. GPS = Global Positioning System. PA = precision agriculture.

Source: USDA, Economic Research Service estimates using data from the Agricultural Resource Management Survey (ARMS) Phase II.

---

[5] Cropland includes all acres that a corn farmer owns plus any acres rented minus any acres rented to others. Sales are not used as a measure of farm size because some production is fed onfarm or stored rather than sold.

Adoption of Precision Agriculture: Capital Expenditures

PA investments include purchases of equipment, installation charges, and the time and effort spent learning how to use and maintain the technologies. These costs are usually not recoverable if use is discontinued (Pindyck, 1988). Compared to some other capital items with active resale markets, like land and tractors, PA equipment is highly specialized with limited resale potential and is usually a sunk cost. There might even be costs to uninstall or discontinue the use of precision equipment. Outsourcing to a custom service provider is another option, also with costs. These factors increase the financial risks of PA adoption. To offset these costs, farmers require a higher expected return to adopt PA technologies (Schimmelpfennig and Ebel, 2016).

Precision Agriculture and Overhead Expenses

Precision agriculture technologies can affect farm finances through operating costs, overhead costs, and changes in yield. PA technologies could reduce operating costs and raise allocated overhead (by substituting capital and labor for operating inputs), or they could do the reverse. PA adoption could raise costs through increased use of inputs (if a need is indicated by mapping), but also raise yields. Since the impact of the technologies on operating costs alone would ignore this fundamental tradeoff between capital and labor, net returns are considered first, beginning with the relative use of hired and unpaid labor.

Adjusted by acreage and bushels harvested, spending on hired labor—which is small relative to the opportunity cost of unpaid labor—is higher for PA adopters in both per-acre and per-bushel terms (table 2). This difference could be associated with a greater need among adopters for specialized information management and field operation skills. Or these technologies may be disproportionately adopted by farms with higher hired-labor costs.

Hired labor costs are presented by both corn planted acreage and bushels harvested to control for the association between adoption rates and farm size. The percentage difference in hired labor costs between adopters and nonadopters of guidance systems is greater per bushel (154 percent) than per acre (73 percent), unlike the other two technologies. One explanation is that these guidance systems are not associated with increased yields, although they may be adopted more often on acreage with less productive land. Also, benefits associated with guidance such as reduced fatigue may be difficult to quantify in terms of yields. Total labor costs (hired and unpaid) per bushel of production are lower for farms that have adopted PA technology, particularly for GPS mapping, but the difference (from nonadopters) for guidance systems is quite small and not statistically significant.

The costs of both unpaid labor (measured as opportunity cost) and machinery also differ between adopters and nonadopters of each of the three PA technologies (table 2). Unpaid labor costs are presented by corn planted acreage and bushels harvested. These costs are an estimate of what the operator could earn by working off the farm.[6] The cost of unpaid labor is lower for adopters in both per-acre and per-bushel terms, while the cost of machinery is higher, especially for guidance

---

[6] The cost of unpaid labor is estimated using whole-farm financial data for all types of farms (from the "CRR" version of ARMS Phase III). ERS analysts use that data to identify the relationships between farm operator attributes—such as age, education, marital status, and location—to their off-farm earnings. Reported attributes of farm operators in the corn version are then used to estimate their predicted off-farm hourly earnings. Phase III data also help to identify the total number of hours worked on the corn enterprise by unpaid workers, which is then used to calculate an opportunity cost per acre or per bushel for unpaid labor.

Table 2
Overhead costs by precision ag practice (2010 corn)

	GPS soil/yield mapping	Guidance system	VRT	Total
	Corn farm spending by technology			*(Average all corn farms)*
Hired labor ($/acre)				$2.45/acre
Adopter	$3.69	$3.50	$2.82	
Non-adopter	$1.91	$2.03	$2.37	
Percent difference	93%**	73%**	19%*	
Hired labor, yield effect (cents/bu)				18 cents/bu
Adopter	21 cents	31 cents	18 cents	
Non-adopter	16 cents	12 cents	17 cents	
Percent difference	31%**	154%**	5%	
Unpaid labor, opportunity cost ($/acre)				$30.29/acre
Adopter	$21.98	$21.19	$24.37	
Non-adopter	$33.88	$33.96	$31.65	
Percent difference	-35%**	-38%**	-23%**	
Unpaid labor, yield effect ($/bu)				$2.00/bu
Adopter	$1.41	$1.71	$1.55	
Non-adopter	$2.27	$2.12	$2.11	
Percent difference	-38%**	-19%*	-27%**	
Total labor, yield effect (hours/bu)				.10 hours/bu
Adopter	.08	.10	.08	
Non-adopter	.12	.11	.11	
Percent difference	-35%**	-9%	-28%**	
Machinery & equipment capital recovery costs (M&E)				$4,335.72
Adopter	$6,404.36	$7,336.84	$5,409.45	
Non-adopter	$3,440.54	$3,122.06	$4,088.45	
Percent difference	86%**	135%**	32%**	
M&E ($/acre)				$79.97/acre
Adopter	$84.70	$85.42	$80.39	
Non-adopter	$77.92	$77.76	$79.87	
Percent difference	9%**	10%**	1%	
M&E, yield effect ($/bu)				$6.17/bu
Adopter	$5.99	$7.53	$6.04	
Non-adopter	$6.25	$5.60	$6.20	
Percent difference	-4%	34%**	-3%	

Legend: Significant difference of means at **99% level, *90%; no star denotes no significant difference.

Notes: Hired labor is any labor paid hourly and includes time spent operating machinery; scouting for weeds, insects and diseases; and other work by hand. Unpaid labor is commonly family labor in some kind of actual or implied partnership with the farm owner. Total labor is the sum of hired (including contract labor) and unpaid labor. Machinery & equipment is measured as capital recovery costs, which provide a more accurate measure of actual farm production service flows from machinery and equipment than do expense amortization and capital depreciation used for tax reporting. GPS = Global Positioning System. VRT = variable-rate technology.

Source: USDA Economic Research Service estimates using data from the Agricultural Resource Management Survey (ARMS) Phase II and Phase III.

systems. This is consistent with machinery that saves labor hours and PA technologies that require capital investments.

Considering labor costs for large and small farms separately (table 3) shows that PA technology adoption on small farms (140-400 acres of cropland) is associated with lower costs of both hired and unpaid labor (although small farms' use of hired labor is typically limited in any event). Large farms (over 1,200 acres) have lower average unpaid labor costs, particularly with guidance systems, but higher hired labor costs with mapping and guidance. Lower unpaid labor costs per acre with guidance adoption on both large and small farms probably indicates fewer hours needed to accomplish standard field operations with the technology, while the higher cost of hired labor on large farms using mapping and guidance could be associated with the expertise and time needed to implement the technologies.

The higher costs of hired labor across technologies raise questions about whether costs for custom services might also be higher. Hired labor and custom services may both substitute for onfarm labor, and custom service providers supply the machinery to perform spraying and harvesting. Custom service costs associated with the three PA technologies are substantially different between large and small farms, partly because providers' charges per acre decline as the number of acres serviced increases. Small farms have higher custom service costs associated with all three PA technologies (table 3), perhaps because small farmers are likely to use custom service providers to create maps for targeting necessary field operations. Small-farm mapping adopters spend over 130 percent more on service work than do nonadopters.

Adoption of PA technology naturally leads to greater expenditures on machinery and equipment as these technologies are capital intensive. Machinery also has a higher expense base (compared to labor costs) and more potential to influence overhead costs. Machinery and equipment costs are measured as capital recovery costs that are derived from useful life estimates, which are better measures of actual service flows than depreciation. These are not the asset values of the machinery, but the annualized value of machinery investment costs.[7] Capital costs for leased machinery are included in these capital recovery costs. Equipment used by custom operators is not included in capital recovery costs. When PA technologies are adopted, machinery and equipment costs per acre are higher on a per-farm basis (table 2).

When machinery/equipment cost differences are measured per bushel, guidance is associated with *higher* machinery overhead costs (table 2). Machinery costs associated with the three PA technologies are different for large and small farms (table 3), as is the case for labor and custom service costs. Small farms have lower machinery/equipment capital service flows and costs with mapping and guidance. The smallest difference in cost between small farm adopters and nonadopters is for VRT (1 percent), but the difference is not statistically significant. VRT is the most capital intensive of the three technologies, with small and large farms having to make comparable investments in machinery for VRT. Large farms exhibit slightly higher costs associated with mapping and guidance.

---

[7] When the impacts of PA technologies on operating profit are estimated empirically, years since tractor replacement is considered because a measure of capital turnover is required for analysis at the field operating level rather than machinery and equipment capital recovery costs, which are from allocated overhead and more relevant to farm net returns.

Table 3

Large and small farm costs by precision ag practice (2010 corn)

	GPS Soil/yield mapping	Guidance system	VRT	Average all large/ small farms
	Average per acre costs for large (over 1,200 cropland acres) and small (140-400 acres) farms by technology			
Large farm hired labor				$4.24/acre
Adopter	$4.58	$4.48	$4.08	(n=416)
Non-adopter	$3.84	$4.01	$4.31	
Percent difference	20%**	12%*	-5%	
Small farm hired labor				76 cents/acre
Adopter	32 cents	28 cents	24 cents	(n=397)
Non-adopter	80 cents	81 cents	81 cents	
Percent difference	-60%**	-65%**	-70%**	
Large farm unpaid labor				$21.12/acre
Adopter	$21.10	$19.38	$20.67	(n=416)
Non-adopter	$21.15	$22.84	$21.30	
Percent difference	-0.3%	-15%*	-3%	
Small farm unpaid labor				$37.72/acre
Adopter	$26.48	$24.79	$29.12	(n=397)
Non-adopter	$38.71	$39.08	$38.61	
Percent difference	-32%**	-37%**	-25%**	
Large farm machinery & equipment				$86.01/acre
Adopter	$87.66	$88.48	$83.88	
Non-adopter	$84.04	$83.58	$86.83	
Percent difference	4%*	6%**	-3%	
Small farm machinery & equipment				$68.85/acre
Adopter	$59.42	$63.97	$68.47	
Non-adopter	$69.69	$69.37	$68.89	
Percent difference	-15%**	-8%**	-1%	
Large farm custom services				$15.91/acre
Adopter	$17.39	$16.49	$18.78	
Non-adopter	$14.14	$15.32	$14.81	
Percent difference	23%**	8%*	27%**	
Small farm custom services				$22.38/acre
Adopter	$47.26	$30.20	$29.67	
Non-adopter	$20.17	$21.56	$21.62	
Percent difference	134%**	40%**	37%**	

Significant difference of means at **99% level, *90%; no star denotes no significant difference.

Notes: Hired labor is any labor paid hourly and includes time spent operating machinery; scouting for weeds, insects, and diseases; and other manual work. Compensation for salaried workers, like some farm managers and other operators, is converted to an hourly wage and included in hired labor. Unpaid labor is commonly family labor in actual or implied partnership with the farm owner. Machinery & equipment dollars per acre is measured as capital recovery costs, which provide a more accurate measure of actual service flows from machinery than expense amortization and capital depreciation. Custom services are provided by fee-for-service companies that can be hired to perform most field operations. Farm size (large or small) is determined from all cropland acres on the farm, while per-acre costs are from corn acres planted. GPS = Global Positioning System. VRT = variable-rate technology.

Source: USDA, Economic Research Service estimates using data from the Agricultural Resource Management Survey (ARMS) Phase II and Phase III.

Input Costs, Precision Agriculture, and Profits From Adoption

In this section, input costs—for fertilizer, pesticide, seed, and fuel—are compared for adopters and nonadopters of precision technologies, both individually (yield monitors, yield mapping, soil data mapping, guidance systems, and VRT) and grouped in six combinations often adopted in tandem (figure 7). The combination of yield monitoring and mapping (YM + Ymap), for example, indicates that the farmer answered yes to both the yield monitoring and yield mapping questions in ARMS. Compared with all other producers who do not use the precision technologies, PA adopters appear to have lower input costs in every category. The differences range from $22 an acre (for yield mapping) to $2 an acre; the small differences for GPS, VRT, and YM+GPS are not statistically significant. Production input costs were tested as explanatory variables in the operating profit model, but none were statistically significant factors in explaining either adoption or profits associated with the technologies, once we controlled for farm attributes.

Farm Operating Profits Associated With Precision Agriculture

On average, farm operating profit of PA adopters was $66 per acre higher than for nonadopters, before controlling for other factors (table 4). However, taking this $66/acre value as the PA gain for farms adopting such technologies would be a mistake because it is an average across different farm sizes, and adopters have much larger farms (480 acres larger, on average). Thus, the higher

Figure 7

Input costs[1] with and without PA technologies

Input production costs per acre (dolllars)

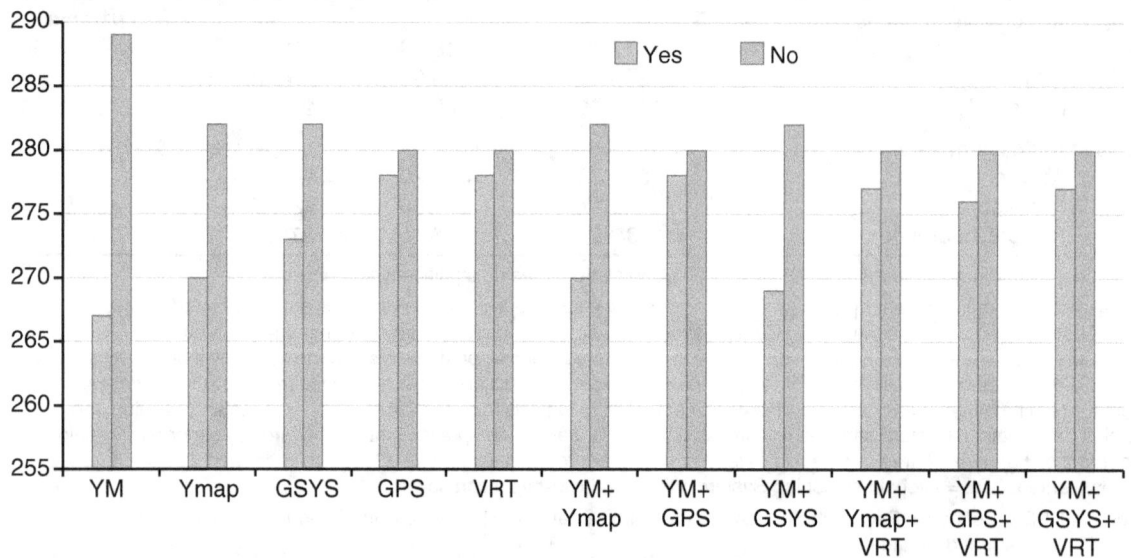

[1]Differences in per-acre costs statistically different at 99% confidence level, except for GPS, VRT, and YM+GPS, which are not significant.
Note: Inputs included in production costs include fertilizer, pesticides, seed, and fuel. GSYS = equipment auto-guidance system. VRT = variable-rate technology. YM = yield monitors. Ymap = yield mapping. GPS = Global Positioning System.
Source: 2010 USDA/Economic Research Service/National Agricultural Statistics Service Agricultural Resource Management Survey (ARMS) of corn producers.

Table 4

Mean value comparisons, subgroups of ARMS data before analysis

Reference group	Mean difference	95% confidence interval[1]	Comparison group
Operating profit of PA[2] adopters (per acre)	$66 higher	$45 to $86 higher	All non-adopters
Farm size of PA adopters	479 acres larger	411 to 546 acres larger	All non-adopters
State PA adoption rates			
Georgia	19 percent lower	34 to 4 percent lower	All other corn States
Iowa	19 percent higher	11 to 27 percent higher	All other corn States
Illinois	16 percent higher	6 to 25 percent higher	All other corn States
Kentucky	0.9 percent lower	15 percent lower to 13 higher	All other corn States
North Carolina	15 percent lower	28 to 3 percent lower	All other corn States
North Dakota	36 percent higher	24 to 48 percent higher	All other corn States
Nebraska	6 percent higher	3 percent lower to 16 higher	All other corn States
New York	28 percent lower	53 to 3 percent lower	All other corn States

Notes: These pairwise comparisons of means assume equal variances between comparison groups.

[1] Indicates 95 percent confidence that mean difference lies in range shown.

[2] PA for "precision agriculture" refers to adoption of Global Positioning System mapping, guidance systems, variable-rate application, or any combination of the three technologies.

Source: USDA, Economic Research Service estimates using data from the Agricultural Resource Management Survey (ARMS) Phase II.

profit figure could reflect economies of scale associated with larger operations, and many factors can influence farm size that might also influence PA adoption (large farms are technology adopters in general). Several Corn Belt States had higher PA adoption rates than the national average. North Dakota's rate is even higher while New York's is much lower.

Profit Impacts of Precision Ag Adoption

Before making empirical estimates of profits from adoption of precision agriculture, we review estimates from the existing literature. Griffin and colleagues (2004) summarize results from many PA studies; of the 87 studies that include corn production,[8] 73 percent report net benefits from PA. For corn studies linked to soil sensing and mapping, benefits exceeded costs 33 percent of the time, compared to 50 percent for corn farms using GPS guidance and 33 percent for farms with yield monitors and VRT. At the time these studies were done (before 2004), soil mapping and yield monitoring/VRT were just recouping costs or worse in two-thirds of the cases studied.

The studies reviewed by Griffin and colleagues indicate that several farm financial variables—including primary occupation of the farm operator, legal organization of the operation, and level of farm financial leverage—help explain the relative profitability of PA technologies. These variables, with 2010 ARMS observations, are shown in table 5.

Erickson and colleagues (2013) surveyed input suppliers about their use of PA technologies. Respondents are typically general managers of Midwestern custom service providers that use PA and dealerships that sell PA equipment and perform custom services. About 70 percent of respondents report that soil sampling with GPS is a profitable service for providers, 50 percent report that their use of guidance is profitable, and over 80 percent of these dealerships report they make a profit on the VRT fertilizer applications they provide. Perhaps guidance makes custom work easier without generating the profits brought in by soil sampling and VRT.

Table 5

Farm operation variables—descriptive statistics for 2010 corn

Variable names	Obs	Mean	Std. dev.	Min	Max
Total net returns ($ per acre)	1,360	85.47	198.95	-891	829
Operating profit ($ per acre)	1,360	343.46	193.32	-483	1,010.8
Farm size (cropland acres)	1,360	1,216.85	1,725.09	6	18,479
Hired labor ($ per acre)	1,360	2.32	6.71	0	87.83
Unpaid labor (opportunity cost per acre)	1,360	35.18	36.46	0	305.58
Machinery & equipment ($/acre of capital recovery costs)	1,360	80.40	38.62	0	524.11
No-till used (adopter=1, not=0)	1,360	0.26	0.44	0	1
Occupation (adopter=1, not=0)	1,360	0.90	0.30	0	1
Legal-org (adopter=1, not=0)	1,360	0.15	0.36	0	1
Debt-assets (ratio in dollars)	1,360	0.14	0.26	0	7.31
New tractor (adopter=1, not=0)	1,360	0.28	0.45	0	1

Note: Definitions for the named variables are discussed in the text.

Source: USDA Economic Research Service estimates using data from the Agricultural Resource Management Survey (ARMS) Phase II and Phase III.

---

[8] Corn production is the most commonly evaluated crop for PA benefits, with 37 percent of the 234 studies in the Griffin survey considering corn. Wheat is the next most commonly evaluated crop, with only a third as many studies that consider PA benefits. Coverage in these PA studies has been presented in percentages above because most studies consider more than one technology/crop combination in corn production.

These results from farm-level PA technology surveys and a separate survey of custom service providers indicate a range of possible influences on the profitability of PA technologies, suggesting a rigorous modeling approach is needed to produce reliable empirical results. The next section discusses requirements for estimating empirical relationships between PA technologies, farm net returns and operating profits, and explanatory variables like farm size and use of hired labor.

Studies Linking PA Technologies to Profit on Corn Farms

Previous empirical work on PA technologies has linked *yield-monitoring maps and GPS soil mapping* to farm profits. Two studies that consider corn and PA profitability issues focus on making the best use of corn yield data collected from combines. Massey and colleagues (2008) mapped yield data using GPS coordinates for a Missouri farm over several years and overlaid actual farm-level costs to convert the yield maps into profitability maps. Field segments with poor topsoil routinely did not make a profit and might be candidates for lower seeding rates, reduced chemical application, or conversion to conservation uses. Similarly, Van Raij and colleagues (2002) argue that GPS soil maps can be used to alter fertilizer applications to suit the soil's characteristics.

Shockley and colleagues (2011) found that *guidance systems* had the potential to influence production practices and increase profits on a hypothetical corn farm in Kentucky. The net results on profit are attributed to better production timing by completing field operations more quickly and from improved field operation accuracy and increases in field capacity for both tractors and self-propelled sprayers.

Empirical studies have linked *variable-rate input application (VRT)* adoption on corn farms to both higher and lower farm profits. While the technical feasibility of pre-programming input applications for site-specific conditions has been well established, the economic feasibility has always been an open question. Several studies focusing on VRT in corn production find mixed results. Thrikawala and colleagues (1999) find that homogeneous fields with low fertility are more profitable under constant nitrogen application rates, which has been standard practice in corn production for many years (Bullock et al., 2009). Only corn fields with sufficient variation in fertility are found to be profitable under variable-rate nitrogen application. Koch and colleagues (2004) found that furrow and center-pivot irrigation for corn production in Colorado were more profitable under variable-rate nitrogen application, but U.S. corn acres are rarely irrigated (the practice was not significant in our results). When Liu and colleagues (2006) included dynamic weather factors in a 3-year panel, they found that variable-rate nitrogen application seldom covered its costs after controlling for site variability.

Even when additional costs, such as operator time spent learning how to implement VRT, are not included, studies find the effect of VRT on profits mixed. Over a quarter of the studies in the Griffin (2004) survey reviewed VRT applications in general, and another 31 percent focus on VRT for nitrogen, seeding, weeds/pests, irrigation, and other nutrient management, with no consensus on profitability.

Predictors of Precision Agriculture Adoption and Operating Profit

Farms adopting PA technology tend to be larger than those that do not. VRT specifically has less variation in use by farm size, but also a lower adoption rate generally. Individual States appear to adopt PA technologies at much different rates (table 4), and this heterogeneity could affect the estimated impacts of PA technologies on operating profit even though it might not have a significant impact on net returns.

Per-acre operating profit for each farm is the gross value of production[9] minus operating (variable production) costs divided by number of planted acres. Variable production costs include all farm input expenditures, including costs for chemicals, custom operations, electricity, fertilizer, fuel, interest on operating capital, irrigation water, pesticides, repairs, and seeds.[10]

Genetically engineered (GMO) seed use was widespread among the farms included in our model, with a 94-percent adoption rate in 2010. GMO seed users on corn farms have twice the adoption rate of nonadopters for mapping and VRT, while guidance adoption is over 10 percentage points higher (table 6). Non-geolocated (non-GPS) soil tests are used by only 29 percent of corn farmers, but soil tests and GPS soil mapping seem to be complementary; users of soil tests use soil and yield mapping at a rate 15 percent higher than farmers who do not conduct soil tests. Non-GPS tests would cover a wider range of nutrients and micro-nutrients but would not be sampled in a grid for a map. Rates for guidance systems and VRT adoption are similarly high among farmers using non-GPS soil tests. This pairing could indicate a general level of comfort with technical production information and a greater affinity for all three PA technologies.

No-till is a form of conservation tillage practice that does not disturb the soil any more than necessary to plant seeds. VRT adopters use slightly more no-till; GPS map and guidance adopters use it slightly less.

Money spent on "all custom services" per acre is higher for adopters of guidance systems and slightly lower for mapping adopters (table 6). When custom fertilization is considered separately from custom pest applications (the two largest uses of custom services), the weak association of PA mapping with custom applications has a clearer explanation. When custom services are used for fertilizer application, mapping is associated with a 12-percent reduction in service cost per acre. When custom services are used for pesticide applications, mapping is associated with a 13-percent *increase* in service cost per acre. Thus, GPS soil/yield mapping is associated mainly with reduced custom fertilization. Guidance systems are associated with higher custom application costs for both fertilizer and pesticides, while VRT use shows much smaller differences in service costs between adopters and nonadopters.

Two additional factors that vary between adopters and nonadopters of PA on corn farm operations are whether crop rotation or irrigation is practiced. In 2010, half of U.S. corn producers planted soy-

---

[9] The gross value of production for corn, for example, is total production (measured in bushels) in a year, valued at the harvest month price. Corn value of production (VOP) differs from corn sales because VOP accounts for corn used onfarm as feed, and because sales during a year could be drawn from production stored from a previous year.

[10] Any production variables that degraded model fit in the models with net returns will be tested in the models that use operating profit.

Table 6

Field operations and precision ag adoption (2010 corn)

	GPS soil/yield mapping	Guidance system	VRT
	Percent of farms adopting each technology		
Genetically engineered seeds planted (94% of farms)			
Adopter	32%	30%	20%
Non-adopter	15%	19%	10%
No-till used (26% of farms)			
Adopter	28%	28%	22%
Non-adopter	32%	29%	18%
Soil tests (29% of farms)			
Multiple nutrients, not GPS-based			
Adopter	42%	40%	28%
Non-adopter	27%	25%	15%
All custom services ($/acre, $18.87 average)			
Adopter	$18.66	$22.05	$20.98
Non-adopter	$18.96	$17.44	$18.39
Percent difference	-2%	26%**	14%**
Custom services ($/acre, $23.84 average) if custom fertilized (50% of farms)			
Adopter	$21.87	$25.20	$24.14
Non-adopter	$24.91	$23.16	$23.72
Percent difference	-12%*	9%	2%
Custom services ($/acre, $28.82 average) if custom pesticides used (54% of farms)			
Adopter	$31.61	$30.98	$29.00
Non-adopter	$27.86	$28.04	$28.78
Percent difference	13%	11%*	1%
Corn – soybean rotation (50% of farms)			
Adopter	35%	31%	22%
Non-adopter	26%	26%	15%
Irrigated corn (6% of farms)			
Adopter	44%	47%	20%
Non-adopter	30%	28%	19%
Stated yield goal at planting (all farms)			
Up to 140 bu/acre	11%	17%	10%
140 – 180 bu/acre	22%	22%	14%
Over 180 bu/acre	49%	42%	28%

Legend: Significant difference of means at **99% level, *90%; no star denotes no significant difference.

Notes: Genetically engineered refers to any seed varieties that have been genetically modified for desirable traits. Custom services are provided by fee-for-service companies that can be hired to perform most field operations. Corn-soybean rotation refers to the practice of alternating years when each crop is grown. GPS = Global Positioning System; VRT = variable-rate technology.

Source: USDA Economic Research Service estimates using data from the Agricultural Resource Management Survey (ARMS) Phase II.

bean the previous year, and this group has higher adoption rates for all three PA technologies than do farms not rotating crops (table 6). Also, GPS mapping and guidance systems are more likely to be used on irrigated corn acres than non-irrigated acres. The problem with irrigation use in examining PA adoption is that only 6 percent of corn acres in the United States are irrigated. Irrigated acres that also are under no-till are beneficiaries of even more guidance (51 percent) and VRT (26 percent) than irrigated acres that are conventionally tilled (27 and 22 percent, respectively).

Farmer-stated goals for corn yields (in bushels per acre) during March-April planting are a good indicator of farmer-perceived profit potential from the land that they farm. Farmers with loftier yield goals have higher PA adoption rates; adoption rates for all three PA technologies nearly double when comparing farmers aiming for 140 to 180 bushels per acre against farmers with a stated goal of over 180 bushels per acre. Descriptive statistics for model variables not discussed previously can be found in table 7.

The strong association between farms with lofty yield-goals and the adoption of all three PA technologies may reflect the geographic distribution of corn farms accustomed to high yields.[11] All three technologies have adoption rates over 30 percent on farms with high yield-goals in Illinois, Indiana, and Iowa, the three States with the highest average yield goals (173-185 bu/acre, as recorded in ARMS). Ohio and Minnesota have the next highest State averages (goals of 164-173 bu/acre); Ohio has a 25-percent adoption rate for mapping and VRT.

Guidance systems are the most common PA technology (between 30 and 60 percent adoption) on high yield-goal farms in all States surveyed by ARMS. Texas has the lowest average yield goal for corn (127 bu/acre), but adoption of all three PA technologies is over 10 percent on farms with both high and low goals for yield. Low yield-goal farms with high rates of guidance system adoption are in Colorado, Kansas, and North Dakota, where average corn farm size exceeds 2,000 acres.

Farmers are well aware of the yield potential on their fields because they are familiar with cash rent prices for land under their control. PA technologies may be seen by adopters as a means to more consistently meet their yield goal. Land that warrants higher yield goals might be expected to have more PA adopters trying to attain those goals (as seen in table 6). This theory will be investigated in the models that control for confounding effects, particularly farm size.

Table 7

Operating profit precision agriculture adoption variables—descriptive statistics (unweighted)

Variable	Obs	Mean	Std. Dev.	Min	Max
Farm size (planted acres)	1,360	1,216.85	1,725.09	6	18,479
GMO seeds	1,360	.92	0.27	0	1
Yield goal	1,360	157.68	39.13	30	300
Soil testing	1,360	.32	0.47	0	1

Notes: GMO seeds have been genetically modified for desirable traits and are also referred to as genetically engineered.
Source: USDA Economic Research Service estimates using data from the Agricultural Resource Management Survey (ARMS) Phase II.

---

[11] Farmer yield goals would usually be formed from several years of actual yields obtained. This would likely be close to the period needed to evaluate capital expenditures for precision agriculture adoption.

Primary occupation, legal organization, and debt-to-assets ratio were investigated as explanatory variables in PA adoption.[12] Farming is the primary occupation for 90 percent of the corn farmers in ARMS, and adoption rates for all three PA technologies are higher when this is the case (20-33 percent adoption across technologies versus 12-13 percent when the primary occupation is other than farming). Farms organized as corporations, estates, or trusts (15 percent of U.S. farms in 2010) use all three PA technologies more than other farms.

Building Empirical Models To Account for Factors in PA Adoption

Up to this point in the report, relationships have been established between PA technology adoption and farm size, labor, machinery, and field operations. These are the key drivers suspected of influencing adoption and net profits. One way to determine how these factors work together or in isolation is to create a robust empirical model to test the impacts of farm size, labor, machinery, and field operation variables on both the previously identified rates of PA adoption and different measures of profit.

One measure of profit, total net returns, includes allocated overhead, covering hired labor expenses plus the opportunity cost of unpaid labor; capital service flows for machinery and equipment; the opportunity cost of land measured as the rental rate, taxes, and insurance; and other general farm overhead. Most investments in PA technologies would fall into the machinery category (see box, "ARMS Data," for variable definitions). The estimated net returns model for farmer i is:

Net returns$_i$ = Function of (overhead expenses$_i$, operating expenses$_i$, and factors affecting precision agriculture technology adoption),

where the technology adoption equation is specified as:

Technology adoption = Function of (Farm size$_i$, yield goals$_i$, overhead expenses$_i$, and operating expenses$_i$)

Overhead expenses include hired labor, unpaid labor, and machinery/equipment capital recovery costs. Operating expenses are determined by field operations defined in table 5, using production inputs that include seed, fertilizer, chemicals, fuel costs, and repairs.

Sample selection is a challenge for modeling profits related to PA technology adoption because the operations with larger profit potential may be more likely to adopt. The treatment-effects model is preferable to other sample selection models when data to estimate profit impacts are available on both adopters (received the technology adoption "treatment") and nonadopters.

The treatment-effects model simultaneously estimates factors explaining technology adoption and those that might also explain profits. There are two key characteristics of this empirical model. The variables explaining technology adoption can be the same as, or similar to, the variables explaining profits without sacrificing model robustness. And there must be some correlation between the adoption and profit sections of the model. If there were no correlation, a separate profit equation could be estimated for each of the PA technologies without any need to consider the impact of adoption. The results from the adoption decision section are used here to estimate the impact of PA adoption on profit, along with other profit-related factors. The statistical appendix describes this procedure in more detail.

---

[12] Of these variables, only primary occupation is significant. These results are shown in table 9 and compared with results obtained for net returns from similar models.

Estimated Results for Total Net Returns

Using net returns as the profit measure, a treatment-effects model is estimated. Larger farms are more likely to adopt mapping and guidance than are smaller farms, but not VRT (table 8, top). Unpaid farm labor, which is an opportunity cost, is negatively associated with adoption. A ready supply of unpaid farm labor may tend to reduce uptake of all three PA technologies. VRT adoption is also negatively associated with the level of service flows (capital recovery costs) from machinery and equipment. This might indicate sunk costs in existing machinery negatively affecting the adoption of capital intensive-VRT.[13]

The profit model—as indicated by the coefficient estimates in the bottom section of table 8—shows positive impacts of PA technology use on net returns, and these are similar across technologies (see table 10 for the percentage impacts on net returns indicated by these coefficients).

Farm size was tested but did not improve the models, which is understandable since it is included in the adoption fitted values. Hired labor (versus unpaid labor for adoption) has a negative impact on net returns associated with adopting guidance systems and VRT; hired labor may be required for periodic

Table 8
Estimated impacts of precision technologies on total net returns

	GPS soil/yield mapping	Guidance system	VRT
	Coefficients estimated separately for each technology (three sets of results are presented)		
Precision technology adoption			
Farm size (cropland acres)	0.055**	0.062**	n.s.
Unpaid labor (opportunity cost)	-0.028**	-0.031**	-0.013**
Machinery & equipment ($/acre of capital recovery costs)	--	--	-0.0077**
Profitability equation			
Precision technology (fitted values from above)	0.752**	0.696**	0.698**
Farm size (cropland acres)	--	--	--
Hired labor ($/acre)	n.s.	-0.013*	-0.011*
Machinery & equipment ($/acre of capital recovery costs)	-0.004**	-0.003**	--

Notes: Estimates **significant at 98% confidence level, or at* 90%. Included variables that are not significant denoted n.s. -- indicates variable not included because it degraded model fit. Costs per acre calculated from acres of corn planted. Farm size determined from total cropland acres, some of which may be used for crops other than corn. GPS = Global Positioning System; VRT = variable-rate technology.

Source: USDA Economic Research Service estimates using data from the Agricultural Resource Management Survey (ARMS) Phase II and III.

---

[13] All of the explanatory variables in the conceptual equations above were tested in this model; those failing to improve model performance are not reported in table 8. The variables that are included are representative of the whole corn farm enterprise; net returns includes all labor costs and farm overhead.

maintenance of PA equipment, which may reduce net returns. Likewise, machinery has a negative impact on net returns associated with mapping and guidance as the additional equipment required for these technologies may be more expensive to buy and maintain. Machinery might be expected to have a negative impact associated with adoption of capital-intensive VRT in particular, but results indicate no impact on net returns. This is not surprising in that machinery and equipment already had a large negative effect on adoption, which is contributing to the fitted values estimate for "precision technology." Unpaid labor degraded model fit in the profit sections of the three PA models, and no other variables were significant in explaining adoption of the PA technologies or net returns.

Operating Profits Versus Net Returns for Computer-Generated Maps

The same type of treatment-effects model used to estimate net returns was also used to estimate operating profit. Here, operators are assumed to only consider factors affecting operating profits when deciding to adopt PA technology (overhead expenses are excluded from the model). For GPS/soil mapping, farm size has the largest positive effect on adoption, soil testing (for non-GPS mapping purposes) the next largest positive effect, and stated yield goal at planting a negative effect. This latter result could indicate that PA technologies are often implemented to help counteract problem fields where a farmer might expect a lower yield because of production challenges. This result contrasts with the simple positive correlation between yield goals and adoption of all three technologies described in table 6, which fails to account for interactions between yield goal and other explanatory variables, particularly farm size and location.

The first row in the lower "profitability equation" section of table 9 uses the information from the adoption section to estimate the impact of PA technology adoption on profit, which is found to be negative. (Interpretation of this coefficient will be discussed below, but after corrections, the coefficient is small and positive—see last row of table 10.) Farm size is positively associated with profit, as is (farmer primary) occupation and new tractor. Years of farming experience[14] has a small negative effect on profitability of GPS mapping, possibly indicating a slight age bias in farmers' ability to use maps profitably. The positive sign on new tractors is consistent with a negative sign on machinery service flows in the net returns estimates for GPS mapping. A higher service flow implies a larger existing stock of machinery and less flexibility (all else equal) to adjust capital stock purchases to profitably accommodate new PA technologies.

Operating Profits Versus Net Returns for Guidance Systems

Operating profit results for guidance systems (from the treatment-effects estimation) are quite similar to mapping (table 9). Farm size has the largest positive effect on adoption, and soil testing is positive and significant. Stated yield goal again has a negative effect on guidance adoption, which is reasonable since GPS guidance is most helpful on fields with variable production conditions (and since farm size is controlled for in the estimates).

Again, farm size has a positive effect on both technology adoption and operating profit. Primary occupation (farming) and new tractor are also positive (lower section of table 9). Years of farming experience degraded model fit with guidance and with variable-rate technology. Primary occupation here (0.20) and hired labor in the net returns guidance estimates (-0.013) have opposite signs, which is plausible if more active managers (with farming as their primary occupation) tend to perform

---

[14] The mean of this variable is 31 years with a standard deviation of almost 14 years.

Table 9

Estimated impacts of precision technologies on operating profit

	GPS soil/yield mapping	Guidance system	VRT
	Coefficients estimated separately for each technology (three sets of results are presented)		
Precision technology adoption			
Farm size (planted acres)	0.44**	0.51**	0.24**
Genetically engineered seeds (yes/no)	--	--	.027**
Stated yield goal at planting (bushels/acre)	-0.70**	-0.77**	-0.53**
Soil tests done (yes/no) Multiple nutrients, not GPS	0.12**	0.14**	0.20**
Profitability equation			
Precision technology (fitted values from above)	-0.79**	-0.85**	-0.78**
Farm size (planted acres)	0.15**	0.17**	0.09**
Farming primary occupation	0.20**	0.20**	0.18**
Years farming experience	-0.06**	--	--
New tractor (since 2005)	0.10**	0.12**	0.10**

Note: ** significant at 95% confidence level or higher. The models estimated and reported here do not have any insignificant variables; -- indicates variable not included because it degraded model fit. Complete details of estimated model results and discussion of econometric approach can be found in the statistical appendix. GPS = Global Positioning System; VRT = variable-rate technology.

Source: USDA, Economic Research Service estimates using data from the Agricultural Resource Management Survey (ARMS) Phase II and III.

more of the hired labor tasks. The positive sign on new tractors for guidance systems could be consistent with the negative sign on machinery service flows in the net returns estimates (table 8), for the same reason as for mapping—probably because of a larger, less flexible, stock of machinery. Overall, the effect of guidance systems on profitability is small and positive (table 10).[15]

Operating Profit Versus Net Returns for Variable-Rate Technology

Adoption results for VRT differ from GPS mapping and guidance results. Farm size is still positive, but with roughly half the estimated impact on adoption and a smaller positive impact on profits (table 9). (Farm size is also not significant in the VRT adoption equation for net returns; table 8). The coefficient for soil tests is positive and larger than for the other two technologies, while the use of GMO seed is significant and positive only for VRT. Stated yield goal is still negative for VRT but the estimate is smaller than for mapping or guidance.

---

[15] In the conversion of estimated technology coefficients (table 9) to percentage impacts (table 10), a correction is made for the fact that farm size appears twice in the estimates (on table 9 farm size is significant in both the adoption and profit sections). Without correction, the impact percentages (table 10) would double-count the impact of farm size (see appendix for further discussion).

Table 10

Impacts of precision technologies on net returns and farm operating profit

	GPS soil/yield mapping	Guidance system	VRT
	Percentage change in profits from adopting		
Net returns (incl. overhead) Impact of precision technology	1.8%	1.5%	1.1%
Operating profit impact including farm size scale effect	2.8%	2.5%	1.1%

Note: Percentages in second row have been corrected for the "farm size" effect discussed in the text, using the estimated correlation between the adoption and profit sections of each treatment model. GPS = Global Positioning System; VRT = variable-rate technology.

Source: USDA, Economic Research Service estimates using data from the Agricultural Resource Management Survey (ARMS) Phase II and III.

In the profit section (with fitted values for VRT adoption), farming as a primary occupation and new tractor are positively associated with operating profit. VRT adoption is negatively associated with profit, but the effect is slightly smaller than for other PA technologies.

These coefficient estimates are translated (table 10) into percentage impacts on operating profits that account for the necessary correlation between the adoption and profit sections of the model, as well as the double-counting of farm size that appears in both the adoption and profit sections of the model (calculations are shown in the appendix.) In the net returns estimates, farm size was not significant in the profit section of the model, which simplified the calculations for table 10. The fact that both sets of estimates using different explanatory and dependent variables produce results that are within 1 percent of each other shows the robustness of the treatment-effects model.

Precision Technology Profit and the Impact of Farm Size

The effect of all three precision technologies on aggregate net returns to U.S. corn farmers is positive in our model. Converting elasticities (table 8) to percentages, the net returns of corn farms that use at least one of these technologies are 1.1 to 1.8 percentage points higher than for corn farms that do not use the technology (table 10). The positive impact on net returns appears to diminish once capital requirements and costs associated with the technologies are modeled. Mapping has lower capital requirements, on average, and the greatest impact on net returns (1.8 percent). VRT, which is usually capital intensive, has a lesser impact on net returns (1.1 percent), with higher capital expenditures likely offsetting some of the gains from using the technology.

Estimating the impact in percentage terms of PA technologies on operating profit requires two steps. First, elasticities (table 9) are converted to percentages (table 10). Second, the percentage estimates are multiplied by average per-acre net returns to correct for the known correlation between the adoption and profit sections of the treatment model; these results, including a scale effect for farm size, are reported in the second row of table 10. In particular, a correction is needed to account for the "double-counting" of farm size in both the adoption and profit equations. This is important because PA adopters had higher profits than nonadopters (table 4) and also had almost 500 acres more to operate, on average, than nonadopters. MacDonald and colleagues (2013) document how average profits on corn farms increased with more acres harvested, up through 2,000 or more acres. Adjustments for farm size produce corrected percentages, which indicate a positive effect of PA technology on profit (1 to 3 percentage points).

The Near Future

This study, in using ARMS data, is the first examination of PA technologies that combines detailed field-level practice information with operator financial information from a nationally representative sample of corn farms. Previous studies have considered the impact of these technologies on experimental or small groups of farms. Results here indicate that similar factors influence the adoption of the PA technologies considered: GPS-based yield and soil mapping (GPS mapping), guidance systems, and variable rate application technologies (VRT). Larger farms had higher rates of adoption of all three technologies, indicating a possible scale effect. Opportunity cost of unpaid labor has an estimated negative effect on technology adoption in the net returns model, as labor may substitute for technology adoption. Soil testing for nutrient deficiencies has a positive effect on adoption in the operating profit model and seems to complement the use of GPS soil and yield mapping, as well as the adoption of guidance and VRT. Stated yield goal has a negative effect on adoption in the operating profit model for all three technologies, suggesting that lower quality land or production inadequacies may dictate the use of these information-based systems when farm size is a control variable.

These adoption results are used to estimate the impact of PA technologies on average net returns and operating profit for U.S. corn farms. The PA technologies increase net returns and operating profit by small percentages. The consistency across the two models suggests that the findings are robust. Farms with more hired labor have lower net returns with guidance and VRT technologies. Larger machinery and equipment capital recovery costs are associated with lower net returns for mapping and guidance. New tractors are associated with higher operating profit with all three PA technologies, consistent with the negative effect of sunk machinery costs that probably restrict options for implementing the guidance/mapping technologies.

Industry publications tout the rapidly growing sophistication of PA technologies and their ease of use. The recent availability of technologies that integrate onboard data platforms with mobile handheld devices and advances in drone/aircraft data collection are examples. Also, farm machinery is increasingly automated, with standard machinery being GPS-ready, and input suppliers are synchronizing data with crop production recommendations.

Cost savings have driven PA adoption in custom applications (Erickson et al., 2013), but the prospect of yield gains or reducing environmental impacts of farming could be driving forces in the future. If PA technologies become easier to implement, they could boost profits for more producers, with environmental benefits. It might be possible to determine if PA technologies tend to influence costs more or less than yields and levels of output. Empirical work such as this could help monitor the inventiveness of the U.S. corn farmer in adapting new technologies to specific needs and help inform future farm policy.

References

Arnade, C., Y. Khatri, D. Schimmelpfennig, C.G. Thirtle, and J. van Zyl. 2000. "A Long-run Profit Function and a Review of the Returns to R&D," *South African Agriculture at the Crossroads: An Empirical Analysis of Efficiency, Technology and Productivity,* J. van Zyl and C.G. Thirtle (eds.). New York, NY: St. Martin's Press.

Bullock, D.S., M.L. Ruffo, D.G. Bullock, and G.A. Bollero. 2009. "The Value of Variable Rate Technology: An Information-Theoretic Approach," *American Journal of Agricultural Economics* 91(1):209-223.

Daberkow, S.G., and W.D. McBride. 2003. "Farm and Operator Characteristics Affecting the Awareness and Adoption of Precision Agriculture Technologies in the U.S." *Precision Agriculture* 4:163-177.

Davey, K.A., and W.H. Furtan. 2008. "Factors That Affect the Adoption Decision of Conservation Tillage in the Prairie Region of Canada," *Canadian Journal of Agricultural Economics* 56(3):257-275.

D'Emden, F.H., R.S. Llewellyn, and M.P. Burton. 2008. "Factors Influencing Adoption of Conservation Tillage in Australian Cropping Regions," *Australian Journal of Agricultural And Resource Economics* 52(2):169-182.

Erickson, B., D. Widmar, and J. Holland. 2013. "Precision Agriculture in 2013," *Crop Life*, 16th Purdue U. Retailer Precision Adoption Survey, W. Lafayette, IN. June. http://agribusiness.purdue. edu/precision-ag-survey.

Fernandez-Cornejo, J., S.G. Daberkow, and W.D. McBride. 2001. "Decomposing the Size Effect on the Adoption of Innovations: Agrobiotechnology and Precision Agriculture," *AgBioForum* 4(2):124-136.

Fernandez-Cornejo, J., C. Klotz-Ingram, and S. Jans. 2002. "Farm-level Effects of Adopting Herbicide-Resistant Soybeans in the U.S.A.," *Journal of Agricultural and Applied Economics* 34(1):149-163.

Greene, W.H. 2007. *Econometric Analysis.* Upper Saddle River, NJ: Prentice Hall, 6th Edition.

Griffin, T.W., J. Lowenberg-DeBoer, D.M. Lambert, J. Peone, T. Payne, and S.G. Daberkow. 2004. *Adoption, Profitability, and Making Better Use of Precision Farming Data.* Staff Paper #04-06, Department of Agricultural Economics, Purdue University, W. Lafayette, IN. June.

Guo, S., and M.W. Fraser. 2010. "Propensity Score Analysis: Statistical Methods and Applications," *Advanced Quantitative Techniques in the Social Sciences*, chapter 4 in *Sample Selection and Related Models*, SAGE Publications, Inc.

Imbens, G.W., and J.M. Wooldridge. 2009. "Recent Developments in the Econometrics of Program Evaluation," *Journal of Economic Literature* 47(1):5-86.

Koch, B., R. Khosla, W.M. Frasier, D.G. Westfall, and D. Inman. 2004. "Economic Feasibility of Variable-Rate Nitrogen Application Utilizing Site-Specific Management Zones," *Agronomy Journal* 96:1572-1580.

Liu, Y., S.M. Swinton, and N.R. Miller. 2006. "Is Site-Specific Yield Response Consistent over Time? Does It Pay?" *American Journal of Agricultural Economics* 88(2):471-483.

MacDonald, J.M., P. Korb, and R. Hoppe. 2013. *Farm Size and the Organization of U.S. Crop Farming*, ERR-152, U.S. Department of Agriculture, Economic Research Service, Aug.

Massey, R.E., D.B. Myers, N.R. Kitchen, and K.A. Sudduth. 2008. "Profitability Maps as an Input for Site-Specific Management Decision Making," *Agronomy Journal* 100(1):52-59.

National Research Council. 1997. *Precision Agriculture in the 21st Century: Geospatial and Information Technologies in Crop Management*, Committee on Assessing Crop Yield: Site-Specific Farming, Information Systems, and Research Opportunities, Washington DC: National Academy Press. http://www.nap.edu/catalog/5491/ precision-agriculture-in-the-21st-century-geospatial-and-information-technologies.

Pindyck, R.S. 1988. "Irreversible Investment, Capacity Choice, and the Value of the Firm," *American Economic Review* 78(5):969-85.

PrecisionAg Buyer's Guide. 2013. Supplement to *CropLife*, Meister Media Worldwide. http://www. precisionag.com.

Schimmelpfennig, D., and R. Ebel. 2011. *On the Doorstep of the Information Age: Recent Adoption of Precision Agriculture*, EIB-80, U.S. Department of Agriculture, Economic Research Service, Aug.

Schimmelpfennig, D., and R. Ebel. 2016. "Sequential Adoption and Cost Savings from Precision Agriculture," *Journal of Agricultural and Resource Economics* 41(1):97-115.

Shockley, J.M., C.R. Dillon, and T.S. Stombaugh. 2011. "A Whole-Farm Analysis of the Influence of Auto-Steer Navigation on Net Returns, Risk and Production Practices," *Journal of Agricultural and Applied Economics* 43(1):57-75.

Tey, Y.S., and M. Brindal. 2012. "Factors Influencing the Adoption of Precision Agricultural Technologies: A Review for Policy Implications," *Precision Agriculture* 13:713-730.

Thrikawala, S., A. Weersink, G. Kachanoski, and G. Fox. 1999. "Economic Feasibility of Variable-Rate Technology for Nitrogen on Corn," *American Journal of Agricultural Economics* 81(4):914-927.

Van Raij, B., H. Cantarella, and J. A. Quaggio. 2002. "Rationale of the Economy of Soil Testing," *Communications in Soil Science and Plant Analysis* 33(15-18):2521-36.

Statistical Appendix

The treatment-effects empirical model used in this report to generate estimated impacts of PA technologies on operating profit is an approach described in Imbens and Wooldridge (2009). The treatment-effects model is appropriate when the empirical data tested have observations on both treated (PA technology adopters) and nontreated (nonadopters) participants. This commonly arises in program evaluation when a subsample of participants is selected to receive a treatment, often a policy instrument or a medical procedure. The model provides consistent and unbiased estimates of the treatment (adoption of PA technology) on an outcome of interest (profit level) because it can control for factors affecting adoption (farm size) that might be confounded with factors explaining profits (also farm size).

The treatment-effects model works by including the yes/no technology adoption choice as an endogenous dummy variable in the adoption section of the model (Greene, 2007). In that section, the technology is the dependent variable; in the profit section, the technology enters as an explanatory variable. The fitted values from the adoption section are used in the profit section to estimate the impact of the technology on profit, while also allowing for the inclusion of other profit-related factors. Both sections are iteratively estimated, simultaneously until parameter convergence in a maximum likelihood estimates (MLE) formulation.

The treatment-effects model is a sample selection-type model, which has been used before to model the adoption of agricultural production technologies (Fernandez-Cornejo et al., 2002). Profit functions like those used by Arnade et al. (2000) did not yield useful results, probably because of the type of structure they impose. Variable selection is accomplished by evaluating model fit, parameter parsimony, and ease of interpretation for different model alignments.

Since R-square coefficients of determination are not available for treatment models, model performance is evaluated using two other criteria: the Akaike and Bayesian Information Criteria (AIC and BIC) (Guo and Fraser, 2010). These diagnostic statistics are measures of combined fit and complexity. Unlike the R-square that is bounded at one (a perfect fit), better models in terms of information criteria have smaller values. The AIC and BIC statistics for the estimated operating profit models are near each other in size and are about 15 percent smaller than the results for net returns, indicating a better fit but not a significant problem for the net returns models. AIC and BIC statistics are reported with the operating profit tables (in this appendix) and in table 8 notes.

Another necessary specification test involves testing that the adoption equation is *not* independent of the equation explaining profit. If these two simultaneously estimated components were found to be independent of each other, it would indicate that profits were not being influenced by the decision to adopt precision agriculture. If this were the case, the maximum likelihood approach would not be necessary and the two sections of the model could be estimated independently. The test statistic for treatment independence comes from the Wald test reported at the bottom of each set of treatment-effect results. This test is a Chi-squared test of rho equal to zero that is rejected at high levels of significance (P values close to zero) in the tables. The Wald test at the top of each table of results is a different test, analogous to an F-test of joint significance of the variables, with the possibility of the estimated coefficients jointly equal to zero rejected.

The end result is that the coefficients reported in appendix tables 1-3 take the available information into account and the coefficient on each PA technology from the fitted values (in the top section of each table) can be reliably interpreted. Additional variables that were tested and found to degrade model fit include input costs like fertilizer, pesticides, seed, labor, and fuel. Variable deletion tests were used to confirm non-significance in relation to other included variables. Farm size was tested as a squared term in addition to farm size levels in the adoption section of the model, and also as an interaction term with the PA technology residuals in the profit section.

Estimated Results for Net Returns

The variables that affect the adoption of GPS yield/soil mapping and guidance systems in appendix table 1 (net returns) are farm size (positively) and unpaid labor (negatively). Unpaid labor is also negative and significant in VRT adoption, while machinery and equipment capital recovery costs are negative and significant, as might be expected because of the capital requirements of VRT. This represents a service flow and not the initial capital outlay, which helps explain the small size of the estimated coefficient on machinery. (These are the same coefficients as summarized in table 8, with interpretation of the fitted values in table 10.) Machinery negatively affects net returns with GPS mapping and guidance, as expected, because of the added capital expenses, but the overall impacts on net returns are positive (table 10). Hired labor has negative and significant impacts on net returns with guidance and VRT. The only remaining variable, "Constant," is the intercept term.

Computer Mapping and Operating Profit

Three variables affect the adoption of GPS yield and soil-mapping technologies (denoted "GPS mapping" in appendix table 2). These are the same coefficients as summarized in table 9. Continuous variables that are not from fitted values, like corn planted acres and yield goal (an indication of yield potential reported by the farmer), are converted to logs for desirable estimation properties so the appendix table 2 estimates can be interpreted as elasticities. Corn planted acres, Soil testing, and Yield goal are significant in explaining mapping adoption (bottom of table), and mapping is significant in explaining operating profit on a per-acre basis.

These results come after controlling for other confounding factors like farm size. After including the impact of scale economies from farm size, the effect of mapping on profits is slightly positive. Profits for adopters of any PA technology are $66 per acre higher (table 4) than for nonadopters when only a simple comparison is made that does not control for other factors influencing profits and PA adoption.

Other variables in addition to GPS mapping (top section of appendix table 2) are significant in explaining profit. One is farming as primary occupation (Farming occupation). Daberkow and McBride (2003) found that full-time farming was significant in both the first and second stages of their PA awareness and adoption empirical model. Years of farming experience has a small negative (but significant) impact in this treatment model. The other variable that is significant is "New tractor since 2005," with farmers who recently bought a new tractor having higher profits, on average, when mapping adoption is considered. This is consistent with new machinery being used to accommodate PA technologies. (It could also indicate merely that farmers with higher profits had the means to buy new tractors.) There were 13 omitted State dummy variables representing the Corn Belt and States with substantial irrigated corn. Profits, holding PA technology adoption constant, are

Appendix table 1

Net returns treatment-effects models—MLEs for all three technologies

	GPS Soil/yield mapping	Guidance system	VRT
Wald chi^2(1) test of independent equations (i.e. rho = 0) (Prob>chi^2)	9.51 (0.002)	11.03 (0.0009)	15.94 (0.0001)
Model Wald chi^2(3) test (pseudo F test)	21.71 (0.0001)	15.00 (0.0018)	14.53 (0.0007)
Log pseudolikelihood	-1,496,188.9	-1,480,560.1	-1,433,492.2
Akaike information criterion (AIC)	2,992,394	2,961,136	2,867,000
Bayesian (Sawa's) information criterion (BIC)	2,992,433	2,961,175	2,867,040
	Coefficients estimated separately for each technology (three sets of results are presented)		
Precision technology adoption			
Farm size (cropland acres)	0.055 (3.38)**	0.062 (4.10)**	n.s.
Unpaid labor (opportunity cost)	-0.028 (-7.23)**	-0.031 (-9.26)**	-0.013 (-2.27)**
Machinery & equipment ($/acre of capital recovery costs)	--	--	-0.0077 (-3.05)**
Profitability equation			
Precision technology (fitted values from above)	0.752 (3.39)**	0.696 (2.99)**	0.698 (3.49)**
Farm size (cropland acres)	--	--	--
Hired labor ($/acre)	n.s.	-0.013 (-1.61)*	-0.011 (-1.52)*
Machinery & equipment ($/acre of capital recovery costs)	-0.004 (-2.99)**	-0.003 (-2.68)**	--
Constant	5.083 (39.14)**	5.082 (41.33)**	4.895 (74.03)**
Sample size (N)	1,014	1,013	1,014

Notes: Estimates ** significant at 98% confidence level, or at * 90%. Z-stats (pseudo t-statistics) in parentheses. Included variables that are not significant denoted n.s. -- indicates variable not included because it degraded model fit. Costs per acre calculated from acres of corn planted. Farm size determined from total cropland acres, some of which may be used for crops other than corn. GPS = Global Positioning System; VRT = variable-rate technology. MLE = maximum likelihood estimate.

Source: USDA Economic Research Service estimates using data from the Agricultural Resource Management Survey (ARMS) Phase II and III.

Treatment-effects model—MLEs for computer maps

Wald test of independent equations (rho = 0): chi^2(1) = 156.00 (Prob > chi^2 = 0.0000)
Model Wald chi^2(11) = 229.29 (Prob > chi^2 = 0.0000)

Log pseudolikelihood = -1,251,444

Akaike information criterion (AIC) = 2,502,924
Bayesian (Sawa's) information criterion (BIC) = 2,503,017

Variables		
Operating profit (per acre)	Corn planted acres	0.149
		(5.70)**
	Farming occupation	0.200
		(2.68)**
	Years farming experience	-0.056
		(-1.86)*
	New tractor since 2005	0.096
		(2.48)*
	GA	-0.506
		(-4.19)**
	IL	-0.148
		(-2.96)**
	KY	-0.541
		(-3.21)**
	NY	0.361
		(2.92)**
	NC	-0.227
		(-2.02)*
	ND	-0.252
		(-3.53)**
	GPS mapping (fitted values from below)	-0.794
		(-12.51)**
	Constant	5.156
		(23.72)**
GPS mapping	Corn planted acres	0.444
		(12.23)**
	GMO seeds	0.105
		(1.15)
	Soil testing	0.118
		(1.84)*
	Yield goal	-0.699
		(14.07)**
Sample size (N)		1,278

Notes: Estimates ** significant at 99% confidence level, or at * 95%. Z-stats (pseudo t-statistics) in parentheses. Costs per acre calculated from acres of corn planted. MLE = maximum likelihood estimate. GMO = genetically modified organism. GPS = Global Positioning System.

Source: USDA, Economic Research Service estimates using data from the Agricultural Resource Management Survey (ARMS) Phase II and III.

negative for the other States (except New York) relative to this group. The same States are significant for all three PA technologies considered.[16]

The explanatory variables in appendix table 2 are similar to those found in other empirical work on the adoption of conservation tillage. Davey and Furtan (2008) included operation acreage, operator age, education, and off-farm income. D'Emden and colleagues (2008) tested farm size, education, and extension variables that could be like "Soil testing." They also tested cropping intensity, which is somewhat like yield goal. We tested deviations of individual farmer yield-goals from their State's average, but the variable degraded model fit whether the State dummies were included or not. Reported no-till use in 2010 corn production from ARMS was tested in the treatment models and degraded model fit in the adoption and profit sections of the treatment model for any of the three PA technologies. The goodness-of-fit information criteria are small for all three PA technology models related to the net returns models and are similar in size across the different technologies.

Guidance Systems and Operating Profit

Two variables affect the adoption of guidance system auto-steering (appendix table 3)—Corn planted acres and Yield goal. These are two of the same three variables that were significant in appendix table 2; as before, guidance is significant in explaining operating profit on a per-acre basis.

Corn planted acres and primary occupation are significant in explaining operating profit, as before, and the coefficients are similar in size to those in appendix table 2. A new tractor also contributes to profit after controlling for guidance system adoption, as it did for GPS mapping. This is consistent with table 2, which showed that capital recovery costs for machinery and equipment ($/acre) were higher when either GPS mapping or guidance systems were adopted. These higher machinery overhead expenses probably represent investments in PA-enabling equipment. Wald test diagnostics are significant at high levels. The goodness-of-fit information criteria are similar in size to those in appendix table 2 but indicate a slightly poorer fit than for the GPS mapping model.

VRT and Operating Profits

Adoption of VRT is affected by all four variables tested (appendix table 4). Corn planted acres and Yield goal are significant and have the same signs as in the other two tables; farm size has a positive effect and yield goal is negative. Soil testing has a positive and significant impact on VRT adoption, and the size of the effect is larger than for the other two technologies. Adoption of another production technology, GMO seeds, is positive and significant and larger than the other impacts measured on adoption.

VRT is significant in explaining operating profit and has a negative effect on a per-acre basis. Corn planted acres and primary occupation are again significant in explaining operating profit with the same signs but the size of the effects is smaller, especially for farm size. The scale effect for VRT in appendix table 4 is affected by this smaller estimate for farm size. A new tractor contributes to profit after controlling for VRT adoption, as it did for the other PA technologies. Wald test diagnostics are

---

[16] As a test, Illinois from the Corn Belt was not included in the omitted group and was not significant in the guidance equation, presumably because it belongs, as we expected, in the omitted group. The omitted group was constructed as contiguous high-intensity corn production States after the dummies for several of these States were not significant. The intention became to control for the other States with idiosyncratic PA adoption practices. States outside the Corn Belt or that were not large irrigated-corn production States had lower profits, holding all else constant. New York is the only exception.

Appendix table 3

Treatment-effects model—MLEs for guidance systems

Wald test of indep. equations (rho = 0): $chi^2(1) = 84.19$ (Prob > $chi^2 = 0.0000$)
Model Wald $chi^2(11) = 214.53$ (Prob > $chi^2 = 0.0000$)

Log pseudolikelihood = -1,258,464

Akaike information criterion (AIC) = 2,516,964
Bayesian (Sawa's) information criterion (BIC) = 2,517,057

Variables		
Operating profit (per acre)	Corn planted acres	0.167
		(5.95)**
	Farming occupation	0.203
		(2.49)*
	Years farming experience	-0.040
		(-1.30)
	New tractor since 2005	0.117
		(2.77)**
	GA	-0.549
		(-3.74)**
	IL	-0.087
		(-1.78)
	KY	-0.469
		(-2.89)**
	NY	0.350
		(3.16)**
	NC	-0.323
		(-2.34)*
	ND	-0.291
		(3.58)**
	Guidance system (fitted values from below)	-0.848
		(-10.80)**
	Constant	4.971
		(20.22)**
Guidance system	Corn planted acres	0.505
		(12.44)**
	GMO seeds	-0.026
		(-0.19)
	Soil testing	0.143
		(1.96)
	Yield goal	-0.765
		(-13.29)**
Sample size (N)		1,278

Notes: Estimates ** significant at 99% confidence level, or at * 95%. Z-stats (pseudo t-statistics) in parenthesis. Costs per acre calculated from acres of corn planted. MLE = maximum likelihood estimate. GMO = genetically modified organism. GPS = Global Positioning System.

Source: USDA Economic Research Service estimates using data from the Agricultural Resource Management Survey (ARMS) Phase II and III.

Appendix table 4

Treatment-effects model – MLEs for variable rate technology

Wald test of independent equations (rho = 0): $chi^2(1)$ = 79.90 (Prob > chi^2 = 0.0000)
Model Wald $chi^2(11)$ = 244.44 (Prob > chi^2 = 0.0000)

Log pseudolikelihood = -1,186,594.7

Akaike information criterion = 2,373,225
Bayesian (Sawa's) information criterion = 2,373,318

Variables		
Operating profit (per acre)	Corn planted acres	0.086
		(3.60)**
	Farming occupation	0.178
		(2.38)*
	Years farming experience	-0.037
		(-1.16)
	New tractor since 2005	0.101
		(2.55)*
	GA	-0.587
		(-4.13)**
	IL	-0.109
		(-2.11)*
	KY	-0.430
		(-2.53)*
	NY	0.338
		(2.61)**
	NC	-0.325
		(2.45)*
	ND	-0.274
		(-5.16)**
	Variable rate application	-0.782
	(fitted values from below)	(-12.11)**
	Constant	5.423
		(24.78)**
Variable rate application	Corn planted acres	0.241
		(6.33)**
	GMO seeds	0.269
		(2.15)*
	Soil testing	0.196
		(2.37)*
	Yield goal	-0.530
		(-10.11)**
Sample size (N)		1,278

Notes: Estimates ** significant at 99% confidence level, or at * 95%. Z-stats (pseudo t-statistics) in parenthesis. Costs per acre calculated from acres of corn planted. MLE = maximum likelihood estimate. GMO = genetically modified organism. GPS = Global Positioning System.

Source: USDA Economic Research Service estimates using data from the Agricultural Resource Management Survey (ARMS) Phase II and III.

both significant at high levels. The goodness-of-fit information criteria (AIC and BIC statistics) are similar in size to appendix tables 1 and 2, indicating that this and the other models fit well. The two criteria are both smallest for the VRT model, suggesting it as the best fitting model.

To calculate actual percentage impacts of the PA technologies on operating profit, allowing for farm size, the estimated coefficients from table 9 are used. Each of the PA technology coefficients estimated from adoption fitted values and reported in the top half (profit section) of the appendix tables are multiplied by average per acre operating profits. These technology coefficients each have to be corrected for the known correlation between the error terms in the two sections of the treatment model (rho in the Wald test at the bottom of the appendix tables). Rho is multiplied by the variance of the error terms from the bivariate normal distribution used to estimate the treatment model, and added to the estimated PA technology coefficient. (The results for each PA technology are shown in the second row of table 10.)

Estimated farm size coefficients from the profit section of the treatment model are used to calculate the average large/small farm size contributions to operating profit, including the PA technology effect. After examining farm sizes for the entire sample, an average small farm is assumed to be 600 acres and an average large farm 2,900 acres. These farm sizes correspond to category break points in table 1. Percent impacts of farm size on operating profit are based on average profit for a farm of this size, not the average profit of the entire sample. The scale impact, calculated as the difference between the profit impact on an average large farm and the profit impact on a small farm, is then added to the PA technology impact calculated in the second row and shown in row three of table 10. After the correction, the percent impact of VRT on net returns and on operating profit is the same. This implies a proportional increase in overhead expenditures associated with VRT that keeps the percent increase in net returns the same as the percent impact on operating profit. If overhead expenditures had remained unchanged with and without VRT, the percent impacts would have risen.

www.ingramcontent.com/pod-product-compliance
Lightning Source LLC
Chambersburg PA
CBHW081753170526
45167CB00009B/4013